MATHS PLUS 5

STAGE 3

MENTALS AND HOMEWORK BOOK

NEW SOUTH WALES SYLLABUS

Harry O'Brien
Greg Purcell

OXFORD
UNIVERSITY PRESS

Contents

Contents

NSW Syllabus Outcomes

Units	1	2	3	4
NUMBER AND ALGEBRA				
MA3-RN-01				
Applies an understanding of place value and the role of zero to represent the properties of numbers				
MA3-RN-02				
Compares and orders decimals up to 3 decimal places				
MA3-RN-03				
Determines percentages of quantities, and finds equivalent fractions and decimals for benchmark percentage values				
MA3-AR-01				
Selects and applies appropriate strategies to solve addition and subtraction problems				
MA3-MR-01				
Selects and applies appropriate strategies to solve multiplication and division problems				
MA3-MR-02				
Constructs and completes number sentences involving multiplicative relations, applying the order of operations to calculations				
MA3-RQF-01				
Compares and orders fractions with denominators of 2, 3, 4, 5, 6, 8 and 10				
MA3-RQF-02				
Determines $\frac{1}{2}$, $\frac{1}{4}$, $\frac{1}{5}$ and $\frac{1}{10}$ of measures and quantities				
MEASUREMENT AND SPACE				
MA3-GM-01				
Locates and describes points on a coordinate plane				
MA3-GM-02				
Selects and uses the appropriate unit and device to measure lengths and distances including perimeters				
MA3-GM-03				
Measures and constructs angles, and identifies the relationships between angles on a straight line and angles at a point				
MA3-2DS-01				
Investigates and classifies two-dimensional shapes, including triangles and quadrilaterals, based on their properties				
MA3-2DS-02				
Selects and uses the appropriate unit to calculate areas, including areas of rectangles				
MA3-2DS-03				
Combines, splits and rearranges shapes to determine the area of parallelograms and triangles				
MA3-3DS-01				
Visualises, sketches and constructs three-dimensional objects, including prisms and pyramids, making connections to two-dimensional representations				
MA3-3DS-02				
Selects and uses the appropriate unit to estimate, measure and calculate volumes and capacities				
MA3-NSM-01				
Selects and uses the appropriate unit and device to measure the masses of objects				
MA3-NSM-02				
Measures and compares duration, using 12- and 24-hour time and am and pm notation				
STATISTICS AND PROBABILITY				
MA3-DATA-01				
Constructs graphs using many-to-one scales				
MA3-DATA-02				
Interprets data displays, including timelines and line graphs				
MA3-CHAN-01				
Conducts chance experiments and quantifies the probability				
MAO-WM-01 Working mathematically				
Develops understanding and fluency in mathematics through exploring and connecting mathematical concepts, choosing and applying mathematical techniques to solve problems, and communicating their thinking and reasoning coherently and clearly				

5	6	7	8	9	10	11	12	13	14	15	16	17	18	19	20	21	22	23	24	25	26	27	28	29	30	31	32	33	34	35
NUMBER AND ALGEBRA																														
MA3-RN-01																														
MA3-RN-02																														
MA3-RN-03																														
MA3-AR-01																														
MA3-MR-01																														
MA3-MR-02																														
MA3-RQF-01																														
MA3-RQF-02																														
MEASUREMENT AND SPACE																														
MA3-GM-01																														
MA3-GM-02																														
MA3-GM-03																														
MA3-2DS-01																														
MA3-2DS-02																														
MA3-2DS-03																														
MA3-3DS-01																														
MA3-3DS-02																														
MA3-NSM-01																														
MA3-NSM-02																														
STATISTICS AND PROBABILITY																														
MA3-DATA-01																														
MA3-DATA-02																														
MA3-CHAN-01																														
MAO-WM-01 Working mathematically																														

UNIT 1

Number and Algebra

SET 1 Basic

1 7 + 5

2 11 – 5

3 8 + 7

4 12 – 6

5 4 × 4

6 5 × 5

7 6 × 6

8 1368c = $ ☐

9 12 ÷ 4

10 20 ÷ 5

11 What is the sum of 16 and 7?

12 What is the product of 5 and 8?

13 Divide 24 by 4.

14 What is the value of 7 in 7326?

15

Peter had $87 but spent $36. How much did he have left?

$ ☐

SET 2 Related multiplication and division facts

Write 4 facts associated with each array.

1

4 × 6 =
× =
÷ =
÷ =

2

3 × 8 =
× =
÷ =
÷ =

3

5 × 7 =
× =
÷ =
÷ =

Space Angles

Match each angle to its name.

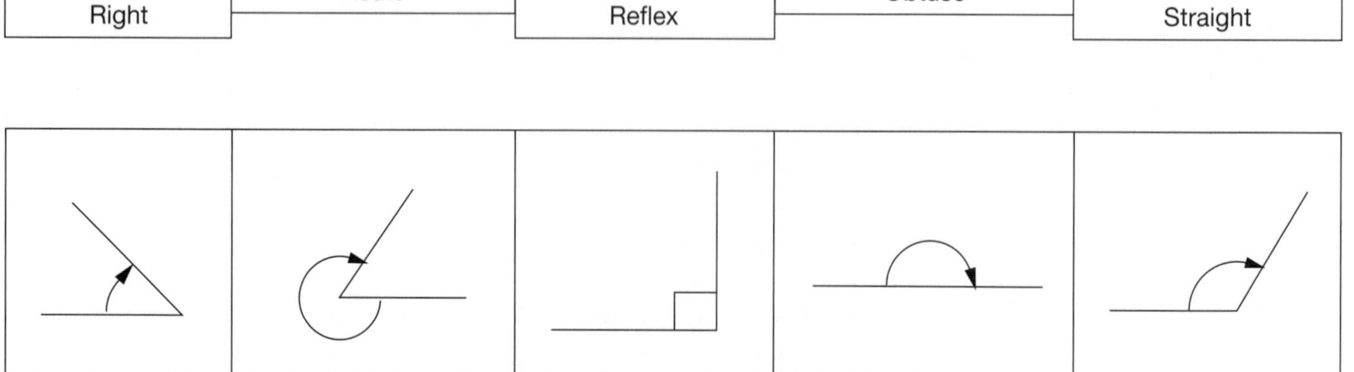

Number and Algebra

SET 3 Addition and subtraction strategies

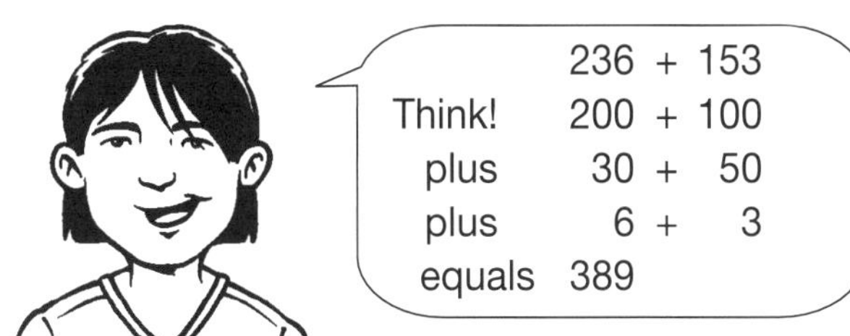

Use the split strategy to answer these additions and subtractions.

1 29 + 54

2 65 – 43

3 58 + 37

4 89 – 27

5 99 + 46

6 375 + 124

7 423 + 254

8 468 – 253

9 227 + 721

10 677 – 205

Working Mathematically

Estimate answers by rounding, then calculate the exact answers.

	Equation	Estimate	Answer
11	49 + 32 =		
12	152 + 149 =		
13	567 – 307 =		
14	317 + 489 =		
15	355 + 246 =		

SET 4 Extension

1 3 m = ☐ cm

2 0.37= ☐ hundredths

3 Write 232 in words.

4 Are 25 and 40 multiples of 5?

5 Share $45 among 5 children.

6 Write the largest number you can, using 1, 4, 2, 6, 8.

7 How many faces has a triangular prism?

8 How much are 4 books at $3.50 each?

9 How many minutes between 12:00 and 1:37?

10 Average of 11, 13, 17 and 23

11 What fraction of June is 15 days?

12 Write the set of factors for 28.

13 1500 g of butter at $3.00 per kilogram

14 $\frac{3}{4}$ of $60

15 Write 26 000 in words.

Measurement Length units

Colour each box to describe the unit of measurement you would use to measure the item.

	Items	mm	cm	m	km
1	length of a ruler				
2	thickness of a coin				
3	length of a road				
4	height of a tree				
5	Olympic marathon				
6	backyard fence				

Number and Algebra

SET 1 Basic

1 16 – 6

2 24 – 6

3 24 + 6

4 18 + 6

5 4 × 8

6 6 × 4

7 7 × 3

8 $32.15 = ☐ c

9 12 ÷ 6

10 15 ÷ 5

11 What is the sum of 35 and 20?

12 What is the product of 20 and 2?

13 Divide 30 by 3.

14 What is the value of 6 in 6312?

15

SET 2 Addition

1 757 plus 213

2 300 + 56 + 620

3 Add 338 and 550.

4 What is the total of 811, 123 and 100?

5

	TH	H	T	O
		3	4	7
		5	2	0
+		1	2	8

6

	TH	H	T	O
		8	1	1
			6	7
+		1	6	0

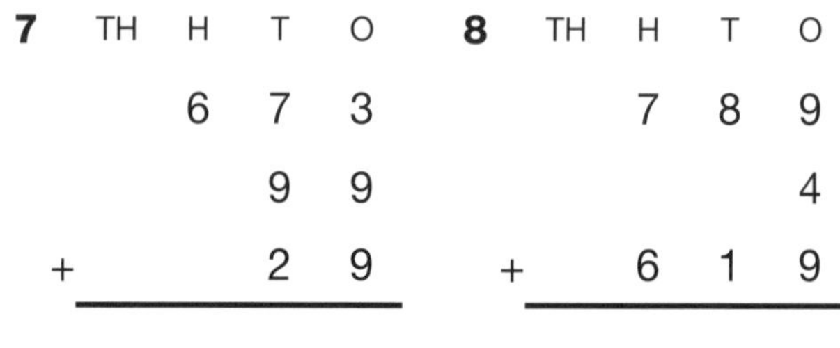

7

	TH	H	T	O
		6	7	3
			9	9
+			2	9

8

	TH	H	T	O
		7	8	9
				4
+		6	1	9

9 Megan bought a microwave for $425, a phone for $365 and a guitar for $117. Megan spent $ ☐.

Space Polygons

Join the dots to make a pentagon, a hexagon and a decagon.

1

5, 4, 1, 3, 2

2

6, 5, 1, 4, 2, 3

3

9, 10, 8, 1, 7, 2, 6, 3, 5, 4

Number and Algebra

SET 3 Thirds and sixths

Write a fraction for each shaded shape.

1 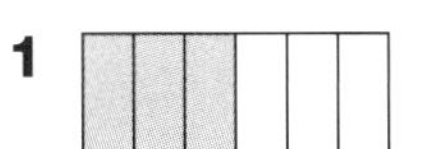 2 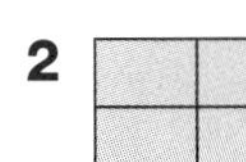 3 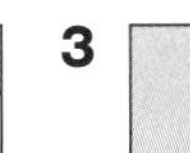 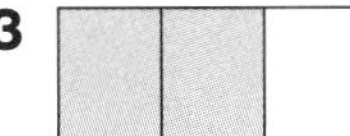

4 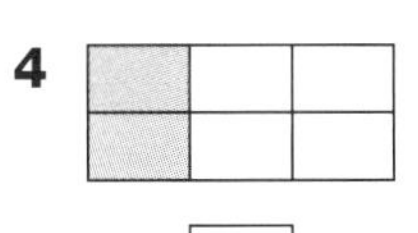 5  6 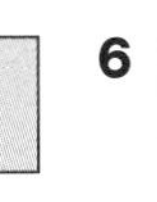 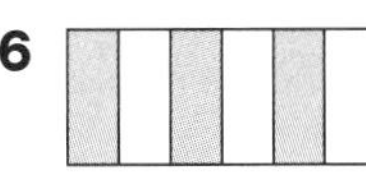

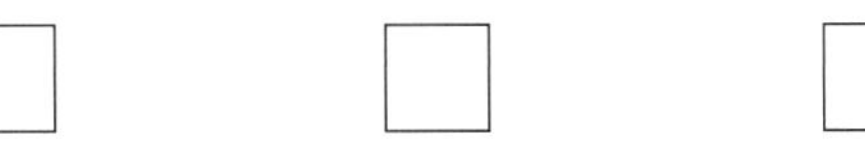

Shade the given fraction of each group.

7 ○ ○ ○ ○ ○ ○ $\frac{4}{6}$

8  △ △ △ $\frac{1}{3}$

9 ▭ ▭ ▭ $\frac{2}{3}$

10 □ □ □ □ □ □ $\frac{2}{6}$

Which is larger:

11 $\frac{1}{3}$ or $\frac{1}{6}$?

12 $\frac{4}{6}$ or $\frac{3}{3}$?

13 $\frac{2}{3}$ or $\frac{5}{6}$?

SET 4 Extension

1 $\frac{3}{5}$ of 30

2 How many days in September, April and June?

3 If 3 kg costs $21, how much would 7 kg cost?

4 36 127 + 3041

5 How many $\frac{1}{3}$s in 2 wholes?

6 How many hours between 9 am and 7 pm?

7 How many edges has a triangular pyramid?

8 232, 257, 282, ☐, ☐

9 Which is the larger: $\frac{1}{2}$ or $\frac{1}{10}$?

10 Write the smallest number you can, using 9, 1, 7, 6, 4.

11 Share $96 between 6 people.

12 How much is $4\frac{1}{2}$ kg of chicken at $4.80 per kilogram?

13 Write the set of factors for 32.

14 $2\frac{1}{4}$ years = ☐ months.

15 Write 39 006 in words.

Measurement Calculating area

Record the length and width of these shapes and then calculate the area.

	Length	Width	Area (L × W)
1			
2			
3			
4			

1

2 cm

4 cm

2 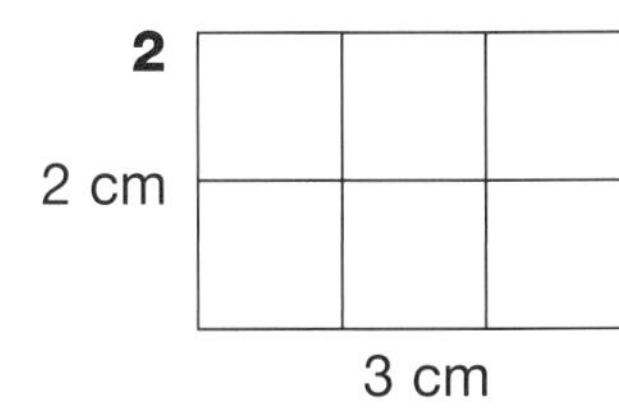

2 cm

3 cm

3 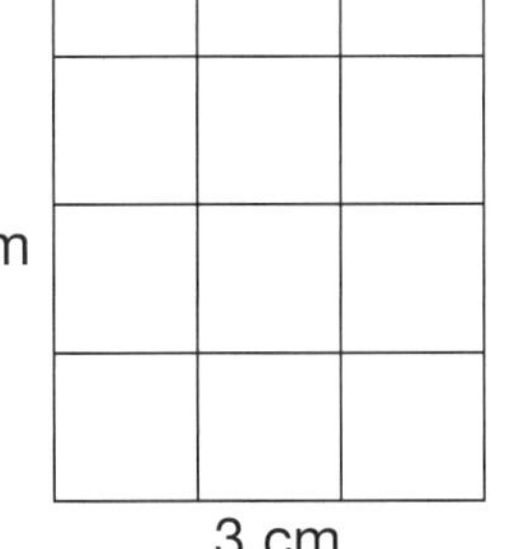

4 cm

3 cm

4 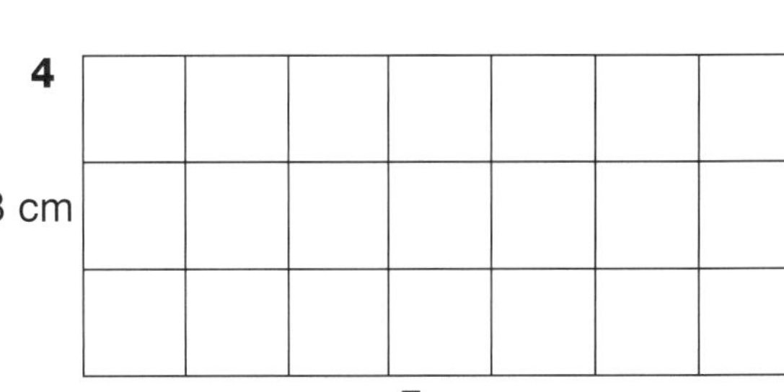

3 cm

7 cm

UNIT 3

Number and Algebra

SET 1 Basic

1 6 + 5

2 9 + 4

3 15 – 4

4 5×8

5 8 – 7

6 9×6

7 7×2

8 $4.21 = ☐ c

9 $18 \div 3$

10 $21 \div 3$

11 What is the product of 6 and 2?

12 What is the sum of 8 and 4?

13 What is the value of 3 in 5382?

14 Divide 21 by 7.

15 

Shaun shared 63 marbles among himself and 6 friends. How many did each child receive?

☐ marbles

SET 2 Subtraction

1 Subtract 123 from 324.

2 $947 take away $123

3 374 minus 163

4 897 take away 456

5 Subtract 354 from 450.

6 374 minus 136

7

	H	T	O
	8	4	7
–	5	2	9

8

	H	T	O
	8	1	1
–	2	6	7

9

	H	T	O
	6	7	9
–		9	6

10

	H	T	O
	7	8	9
–	4	9	4

11 The hot dog stand at the school fete collected $838. If the ingredients cost $576, how much profit did the stand make? 

Statistics and Probability Using fractions

A bag contains 20 coloured balls. Draw a line to match the fractional chance of a single ball being drawn out of the bag for each colour.

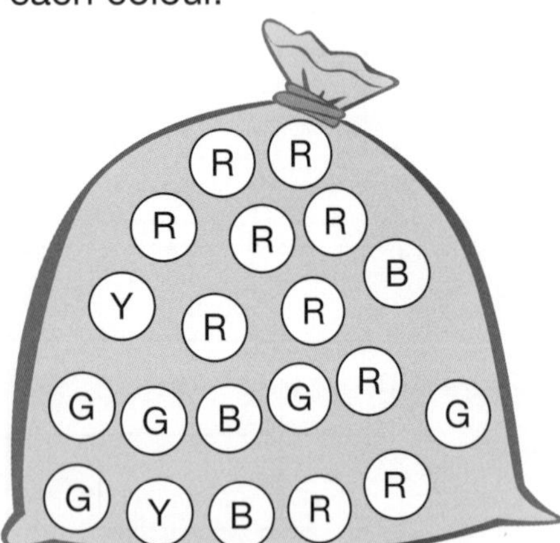

1 

2 Red 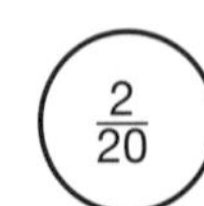

3 Yellow 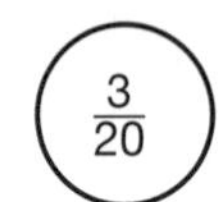

4 Blue 

Number and Algebra

SET 3 Division

Solve the secret code.

1 36 ÷ 3
2 56 ÷ 4
3 92 ÷ 4
4 81 ÷ 3
5 54 ÷ 2
6 91 ÷ 7
7 96 ÷ 6
8 85 ÷ 5
9 102 ÷ 6
10 108 ÷ 4

L	O	R	P	Y	E	K	A	X	C
27	17	12	41	13	14	36	23	92	16

1	2	3	4	5	6

7	8	9	10

Working Mathematically

Write 2 division facts from each multiplication fact.

11 3 × 9 = 27 ______ ______
12 7 × 5 = 35 ______ ______
13 5 × 8 = 40 ______ ______
14 6 × 7 = 42 ______ ______
15 7 × 9 = 63 ______ ______
16 8 × 9 = 72 ______ ______

SET 4 Extension

1 ☐ ÷ 7 = 40
2 How much are 6 watches at $48 each?
3 435 cm – 40 cm
4 Tenths in $2\frac{1}{2}$
5 Average of 150, 200, 100 and 130
6 Double 237.
7 What number is halfway between 180 and 198?
8 What is half of $72.96?
9 $2.30 – 75c
10 Divide 240 by 3.
11 What is the difference between 144 and 40?
12 How much change did I receive from $20 if I spent $7.35?
13 If 4 cost $12, how much would 9 cost?
14 What is the sum of 3564 and 2035?
15 Centimetres in 3.7 m
16 Complete the grid.

18	9	36	27		72	45	
2			3	6		5	7

Measurement Cubic centimetres

Centicubes have been used to build these prisms. Record the volume of each prism.
Colour the object that has a different volume from the rest of the group.

1
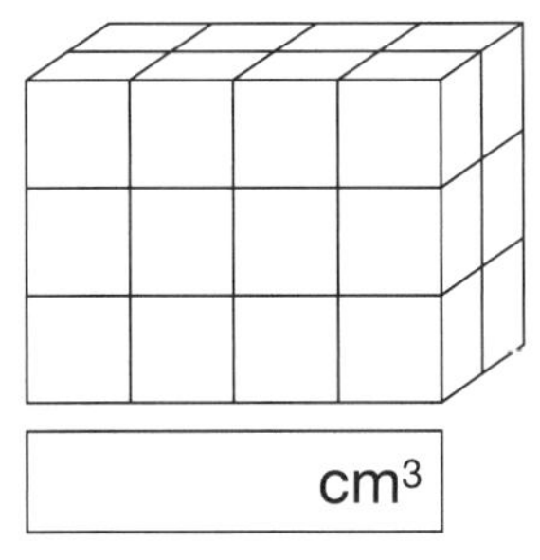
____ cm³

2
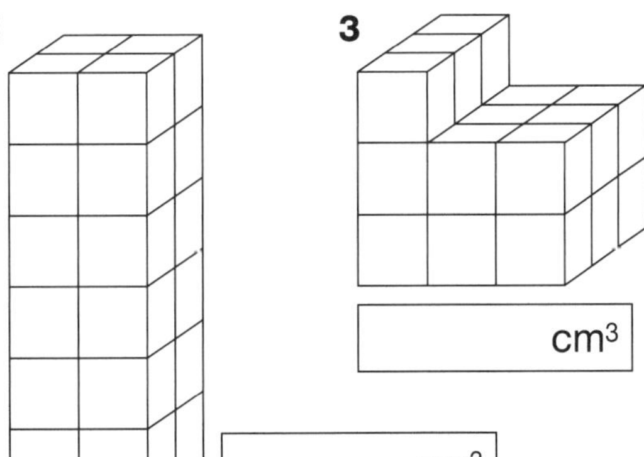
____ cm³

3
____ cm³

4
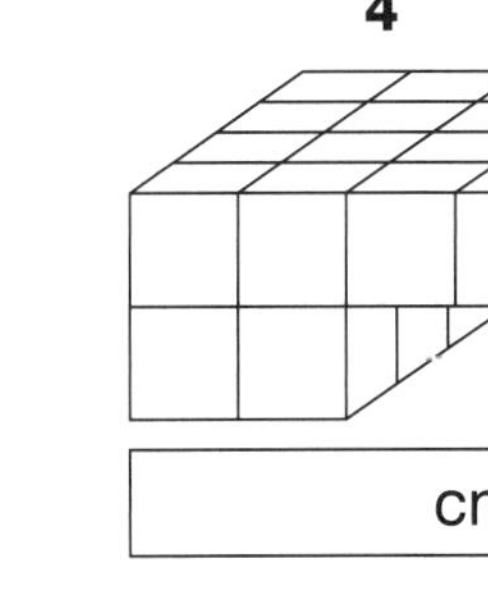
____ cm³

UNIT 4

Number and Algebra

SET 1 Basic

1 7 + 4

2 8 + 9

3 19 – 5

4 11 – 8

5 7 × 2

6 6 × 4

7 8 × 3

8 $4.68 = ☐ c

9 24 ÷ 4

10 30 ÷ 3

11 What is the product of 7 and 4?

12 What is the sum of 8 and 6?

13 What is the value of 4 in 8416?

14 Divide 56 by 8.

15 

Ravi had 26 baseball cards and 52 football cards. How many cards did he have altogether?

☐ cards

SET 2 Addition and subtraction strategies

1 6 + 9

2 60 + 90

3 600 + 900

4 70 + 80

5 700 – 500

6 30 + 90

7 200 + 600

8 700 + 900

9 80 – 30

10 900 – 300

58 – 31
Think 58 minus 30 minus 1 equals 27

Mentally add or subtract these numbers using the compensation strategy.

11 37 + 49

12 53 – 29

13 68 + 19

14 77 – 39

15 55 + 32

16 54 + 29

17 86 – 71

18 38 + 38

Space Coordinates

Use the coordinates to record the position of each shape on the number plane.

1 Triangle ____________

2 Rectangle ____________

3 Circle ____________

4 Pentagon ____________

5 Square ____________

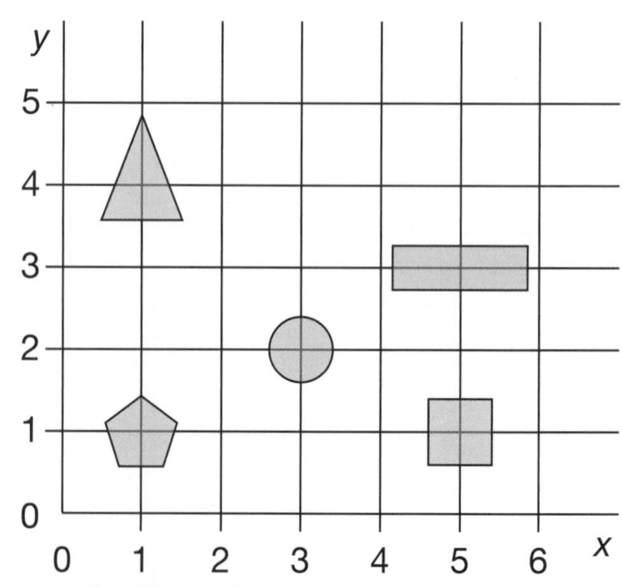

Number and Algebra

SET 3 Place value

Write the place value of each **bold** number.

	Number	Place value
1	**2** 306 504	millions
2	41**7** 653	
3	1 358 6**0**4	
4	12 **2**65 814	
5	11 185 **4**63	
6	8 6**6**1 005	
7	**9** 400 689	
8	1 000 10**1**	
9	27 **8**84 611	
10	2 001 **6**54	

11 Order the numbers from smallest to largest.
659 803, 1 484 003, 1 030 584

____

12 Order the numbers from smallest to largest.
12 694 805, 6 798 612, 9 365 804

____

13 Write 2 605 412 in words.

____

____

SET 4 Extension

1 (3 + 6) × 8

2 If 3 lollies cost 90c, how much would 20 cost?

3 \$867 – \$342

4 Which is larger: 0.7 or 25 hundredths?

5 How many grams in 2.7 kg?

6 How many minutes from 10 am to 2 pm?

7 Double 16 and add 83.

8 How many 5c lollies can be bought with \$1.95?

9 How much are 5 plants at \$9.50 each?

10 Divide 336 by 6.

11 What is the difference between 155 and 36?

12 How much is $3\frac{1}{2}$ kg of potatoes at \$3.50 per kg?

13 10% of 300

14 How many tens in 537?

15 One quarter of 260

16 1, 1, 2, 3, 5, 8, 13, ____ ____ ____ ____

Leonardo Fibonacci discovered this number pattern. Write the next four Fibonacci numbers.

Working Mathematically

Measurement Estimating mass

Colour the masses which could balance a small boy on a seesaw.

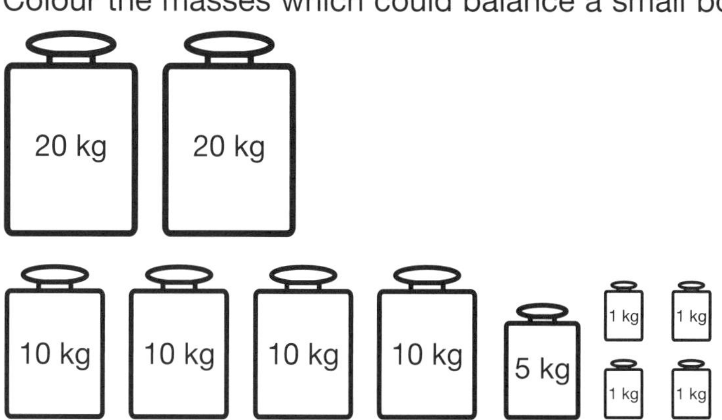

UNIT 5

Number and Algebra

SET 1 Basic

1 6 + 7

2 8 + 5

3 9 − 8

4 10 − 7

5 5 × 8

6 6 × 9

7 4 × 8

8 $6.71 = ☐ c

9 54 ÷ 9

10 63 ÷ 9

11 What is the product of 8 and 7?

12 What is the sum of 6 and 13?

13 Divide 40 by 8.

14 What is the value of 3 in 435?

15 

How many weeks will it take George to save $28 to pay for a CD if he saves $4 per week? ☐ weeks

SET 2 Multiplication strategies

Complete the facts.

		×4	×40
1	4		
2	6		
3	8		
4	10		
5	9		
6	7		

		×6	×60
7	4		
8	6		
9	8		
10	10		
11	9		
12	7		

13 How much are 5 dresses?

14 How much are 7 shorts?

15 How much are 9 T-shirts?

16 How much are 8 books and a dress?

Statistics and Probability Chance experiment

Red, green, pink and yellow marbles are placed in a bag.

1 Which colour marble has the most chance of being pulled out of the bag?

2 Which colour marble has the least chance?

3 Is it more likely that red will be picked than green?

4 Is it more likely that green will be picked than yellow?

5 Which colour has an even chance of being picked from the bag?

Number and Algebra

SET 3 Problem solving

Distance from Sydney

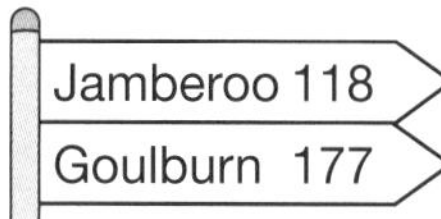

Hat Head 459
Gosford 75

Find the difference in the distance from Sydney of:

1. Goulburn and Jamberoo
2. Goulburn and Gosford
3. Hat Head and Jamberoo
4. Hat Head and Gosford
5. Hat Head and Goulburn
6. Jamberoo and Gosford.

7 When Stephen left home he had \$125. When he came home he had \$36 but thought he should have had \$56.

These are the amounts he spent: \$19, \$20, \$20, \$30.

Did Stephen lose \$20 or not? ________

Working Mathematically

SET 4 Extension

1. \$26.54 = ☐ c
2. $5^2 + 1325$
3. Write the largest number you can, using 3, 7, 6, 9, 1.
4. \$9.50 x 10
5. How many days in autumn?
6. If this Friday is the 13th, what is the date of the following Monday?
7. How much is 250 g at \$8.60 per kilogram?
8. How many 25 cm lengths in $4\frac{1}{2}$ m?
9. 37 more than 1986
10. If 7 cost \$1.40, how much would 9 cost?
11. Are 27 and 35 multiples of 4?
12. How many 500 g bags are needed to fill a 12 kg bag?
13. How much is 5 kg at \$7.80 a kilogram?
14. What is the difference between 240 and 137?
15. $(4 \times 10\,000) + (6 \times 1000) + (3 \times 100) + (9 \times 10) + 8$
16. What type of angle would be formed if the hands on a clock read 1 o'clock?

Space Classifying 3D objects

Name each object.

1

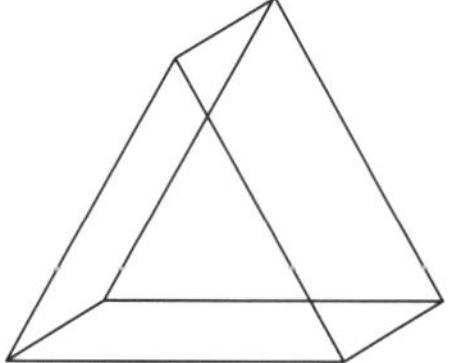

____

2

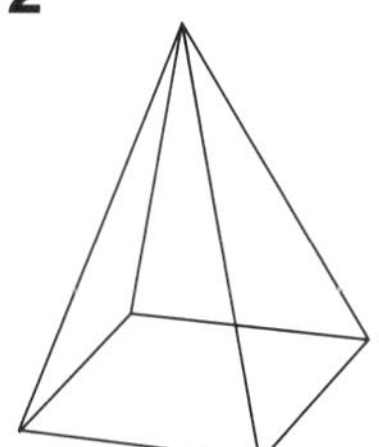

____

3

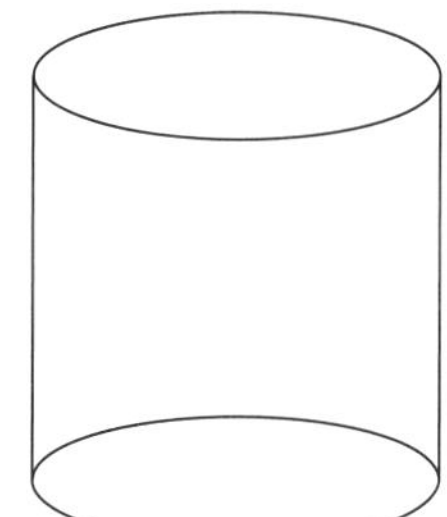

____

4

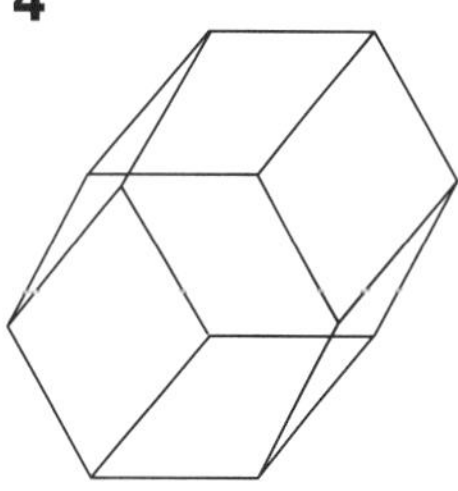

____

5

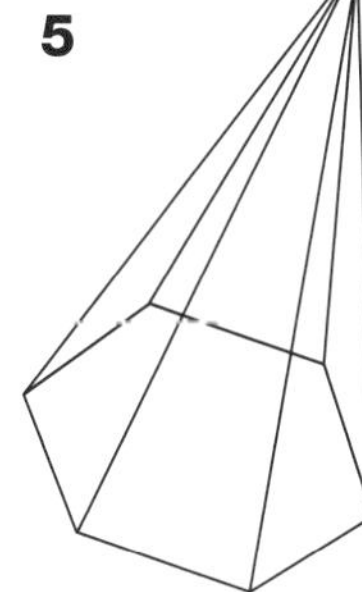

____

UNIT 6

Number and Algebra

SET 1 Basic

1 7 + 5

2 9 + 4

3 8 − 3

4 9 − 7

5 6 × 4

6 8 × 5

7 9 × 3

8 $6.80 = ☐ c

9 49 ÷ 7

10 48 ÷ 6

11 What is the product of 9 and 4?

12 What is the sum of 9 and 11?

13 Divide 42 by 6.

14 What is the value of 4 in 964?

15

A bag of lollies cost 25c. How much would 10 bags cost?

$ ☐

SET 2 Division

1 $\begin{array}{r} 18 \text{ r}2 \\ 3\overline{)\,56} \end{array}$

2 $4\overline{)\,66}$

3 $5\overline{)\,72}$

4 $5\overline{)\,78}$

5 $6\overline{)\,94}$

6 $7\overline{)\,88}$

7 $6\overline{)\,69}$

8 $7\overline{)\,87}$

9 $8\overline{)\,97}$

10 $5\overline{)\,76}$

11 $7\overline{)\,99}$

12 $4\overline{)\,98}$

13 Share 85 lollies among 6 children.

14 Share 37 balls among 4 groups.

15 Share 55 books among 7 children.

16 Share $95 among 5 boys.

17 Share 89 pencils among 8 girls.

Working Mathematically

18 Sascha bought 4 movie tickets and paid for them with a $50 note and received $14 change. How much was each movie ticket?

Statistics and Probability Column graphs

Mr Evans created this column graph to record how many of each type of birthday card his shop sells.

1 Which price card is most popular?

2 Do many people pay $10 for a birthday card?

3 Which cards were not sold at all?

4 What was the total amount of money received from the sale of the $4, $5 and $6 cards?

Birthday card sales

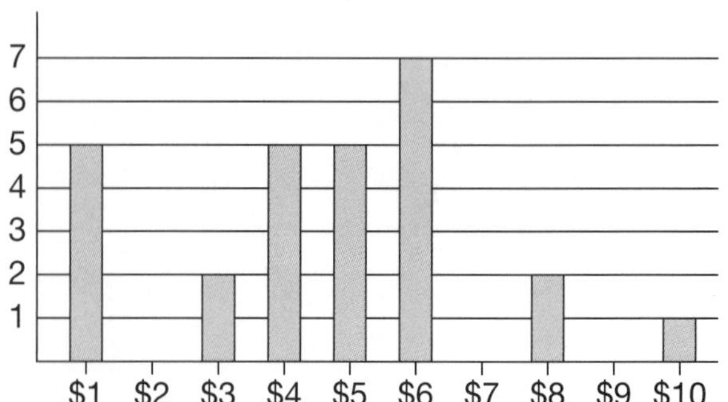

Number and Algebra

SET 3 Number patterns

1 Complete the grid to describe how many sticks were used to construct the pattern.

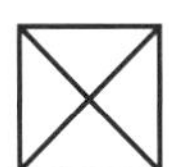

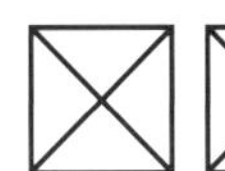
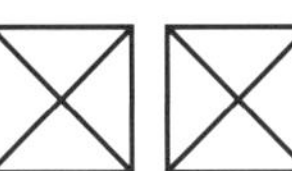

Squares	1	2	3	4	5	6	7
Sticks	6	12					

2 There is one number that is under 50 that is both a square and triangular number. What is it?

Complete the number patterns.

3

23	27	31	35			

4

2	7	12	17			

5

59	56	53	50			

6

1	2	4	8			

SET 4 Extension

1 Estimate an answer to 345 plus 901.

2 $6^2 + 434$

3 $8\frac{1}{2}$ kg = ☐ g

4 $(72 \div 8) + 7$

5 Double 21 334

6 If this Tuesday is the 21st, what is the date of the following Saturday?

7 \$6.70 – \$4.80

8 How many sides do 9 octagons have?

9 What is the product of (28 – 7) and 6?

10 How much are 5 combs at \$4.20 each?

11 0.73 = ☐ tenths + ☐ hundredths

12 $\frac{3}{5}$ of \$80

13 How much is 9 kg of sausages at \$2.15 per kg?

14 What is the difference between 9999 and 999?

15 $(5 \times 10\,000) + (6 \times 1000) + (9 \times 100) + (6 \times 10) + 9$

16 If the large hand is on 11 and the small hand almost on the 9, what time is it?

Measurement am and pm

Express the times shown on the analog clocks using **am** and **pm**.

1 
morning
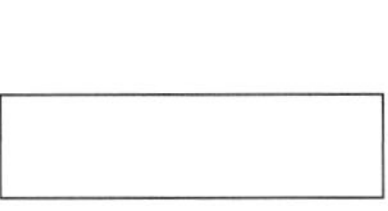

2 
morning

3 
afternoon

4 
evening
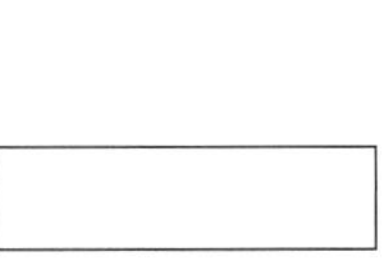

5 
evening
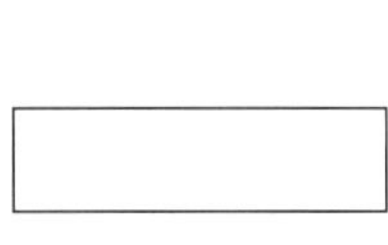

UNIT 7

Number and Algebra

SET 1 Basic

1 6×7

2 8×5

3 $25 - 13$

4 $43 - 5$

5 $18 + 9$

6 $16 + 5$

7 $25 \div 5$

8 $36 \div 6$

9 20, 40, 60, ☐

10 What is the product of 7 and 5?

11 Divide 35 by 7.

12 How many minutes in 1 hour?

13 How much are 2 pears at 40c each?

14 Write the value of 9 in 93 562?

15 How many 50c packets of gum can Theo buy for $5? ☐

SET 2 Multiplication

1

	H	T	O
		2	7
×			4

2

	H	T	O
		6	5
×			6

3

	H	T	O
		3	7
×			8

4

	H	T	O
		4	0
×			5

5

	H	T	O
		6	7
×			6

6 How many players are there if each of the 7 teams has 21 registered members? 

7 How many pupils are there at Cook High if there are 95 in each of the 6 grades? 

8 How much will it cost to buy 9 DVD players at $99 each? 

Space Angles

Use a protractor to measure these angles.

1 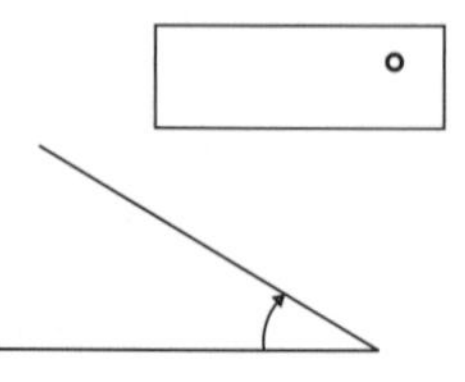

2 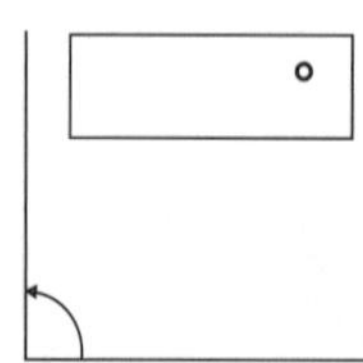

3 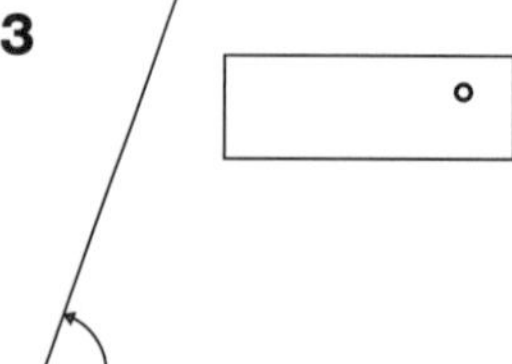

4 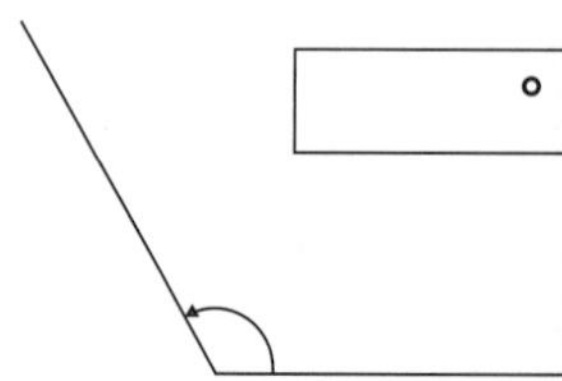

5 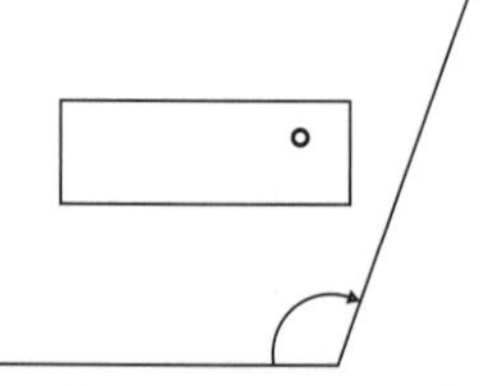

6 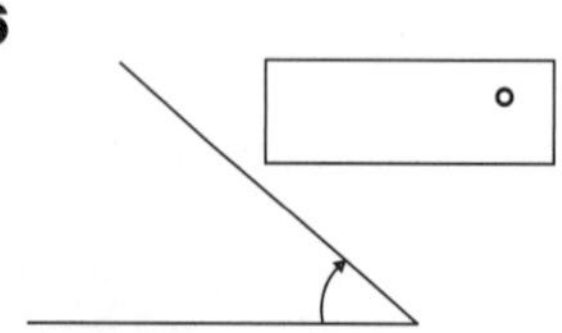

7 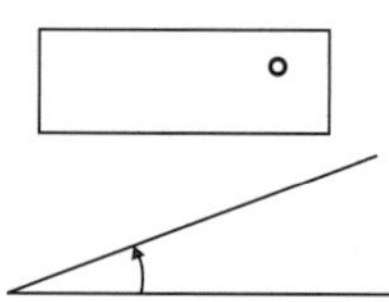

Number and Algebra

SET 3 Fractions

Shade $\frac{1}{2}$ of each shape and write an equivalent fraction for $\frac{1}{2}$.

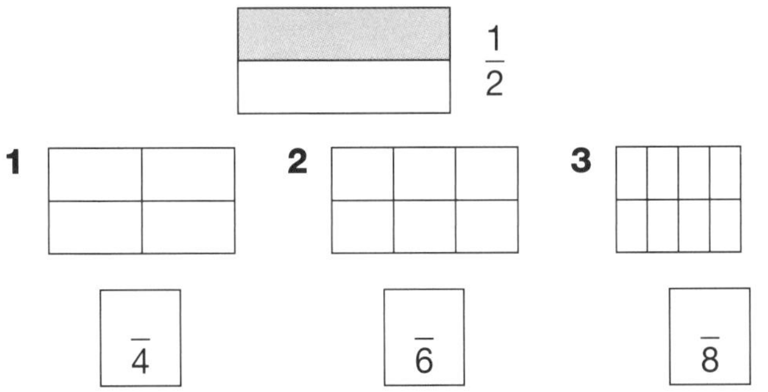

Shade $\frac{2}{3}$ of each shape and write an equivalent fraction for $\frac{2}{3}$.

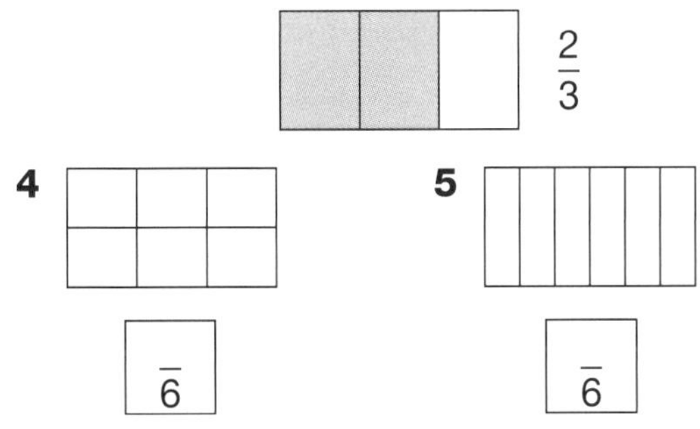

Shade the pair of equivalent fractions in each group.

6	$\frac{1}{3}$	$\frac{1}{2}$	$\frac{2}{6}$	**7**	$\frac{2}{3}$	$\frac{3}{4}$	$\frac{4}{6}$
8	$\frac{1}{2}$	$\frac{1}{4}$	$\frac{4}{8}$	**9**	$\frac{2}{4}$	$\frac{1}{2}$	$\frac{1}{3}$
10	$\frac{1}{4}$	$\frac{4}{10}$	$\frac{2}{5}$	**11**	$\frac{1}{6}$	$\frac{3}{4}$	$\frac{6}{8}$

SET 4 Extension

1 Write $\frac{81}{100}$ as a decimal.

2 How many $\frac{1}{4}$s in $4\frac{3}{4}$?

3 $4^2 + 3$

4 Average 80, 60, 70 and 110.

5 How many 50 cm lengths make 10 m?

6 Write these numbers in ascending order: 4127, 2356, 26 007.

7 How many 250 g packets make 10 kg?

8 Are 45 and 18 multiples of 9?

9 How much are 5 kg of potatoes at $3.50 per kilogram?

10 What is the sum of odd numbers between 14 and 18?

11 What number is halfway between 1420 and 1430?

12 Write the set of factors of 40.

13 420 minutes = ☐ hours

14 Write $2\frac{1}{2}$ as a decimal.

Working Mathematically

15 In our class there are 5 girls to every 3 boys. How many girls are there if there are 32 children in the class altogether?

Measurement Metres to kilometres

Calculate the perimeter of these paddocks in metres and kilometres.

1

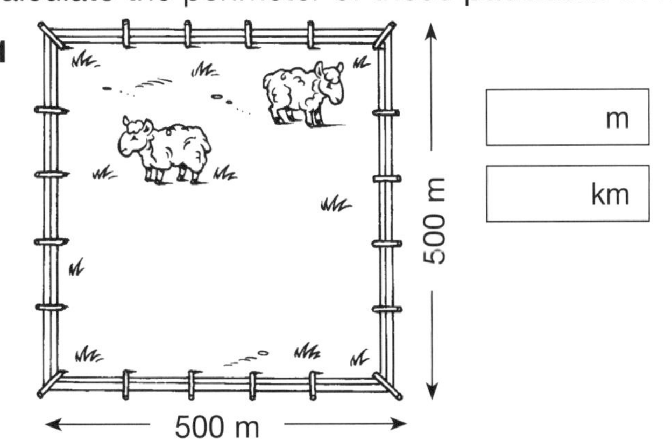

2

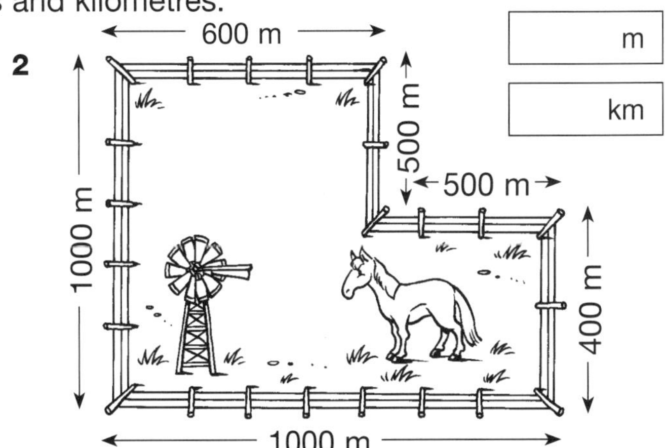

UNIT 8

Number and Algebra

SET 1 Basic

1 25 ÷ 5

2 32 ÷ 4

3 16 + 5

4 27 + 6

5 32 − 5

6 4^2

7 What is the value of 6 in 2364?

8 7 × 4

9 6 × 5

10 \$26.07 = ☐ c

11 What is the sum of 18 and 17?

12 What is the product of 9 and 7?

13 Divide 32 by 8.

14 What is the difference between 28 and 13?

15

Billy bought 3 dozen eggs but dropped them. If 24 were broken, how many were unbroken?

☐ eggs

SET 2 4-digit addition

1

	TH	H	T	O
	3	0	6	4
+	2	7	7	5

2

	TH	H	T	O
	6	5	0	8
+	2	9	5	7

3

	TH	H	T	O
	4	9	0	8
+	3	7	2	5

4

	TH	H	T	O
	8	9	6	5
+	1	0	2	9

5

	TH	H	T	O
	3	0	9	6
+	5	9	6	7

6

	TH	H	T	O
	4	7	1	3
+	4	0	0	9

7 Add 1035, 50 and 200.

8 3000 + 450 + 230

9 What is the total of 170, 230 and 500?

10 800 + 400 + 600

11 What is 2300 more than 3100?

12 What is the sum of 5550 and 1150?

Space Drawing prisms

Repeat each prism on the isometric dot paper. Each one has been started for you.

1

2

3

Number and Algebra

SET 3 Missing numbers

1 9 + ☐ = 18

2 25 + ☐ = 41

3 40 − ☐ = 26

4 75 − ☐ = 54

5 7 × ☐ = 42

6 9 × ☐ = 54

7 48 ÷ ☐ = 8

8 32 ÷ ☐ = 4

9 45 ÷ ☐ = 9

10 15 × ☐ = 45

Working Mathematically

Solve these missing number sentences. Each shape has the same value in every question, for example, the triangle is always equal to 7.

11 ☐ + △(7) = 17

12 △ + ○ = 19

13 ○ + ☐ = ____

14 ☐ + ○ = ⬡ = 25

15 ⬡ × ○ + ☐ = ____

16 (☐ − ⬡) × ○ = ____

SET 4 Extension

1 Add the even numbers between 13 and 17.

2 If 7 kg costs $35, how much would 9 kg cost?

3 $\frac{1}{10}$ = 0.1. True or false?

4 Which is larger: 1.0 or 0.1?

5 How many 0.75 m lengths in 3 m?

6 How many quarters in $5\frac{1}{2}$?

7 How many millimetres in 16.5 cm?

8 Write the smallest number you can, using 4, 2, 3, 6, 7.

9 What shape are the sides of a triangular prism?

10 Write one hundred and fifty-four in numerals.

11 Are 28 and 60 multiples of 9?

12 What time is 13 hours after 3 pm?

13 What is the price of 250 g of nuts at $8 per kilogram?

14 Write 25 minutes past 7 am in digital form.

15 Write 30 027 in words.

☐

16 How many legs on 23 children and 17 horses?

Statistics and Probability Column graphs

The following numbers of home runs were hit by these baseball players.

Name	Home runs
Bart	卌 卌 II
Betty	卌 卌
Darsh	卌 卌 卌 I
Yindi	卌 卌 IIII
Luke	卌 卌

Create a column graph from the home run data.

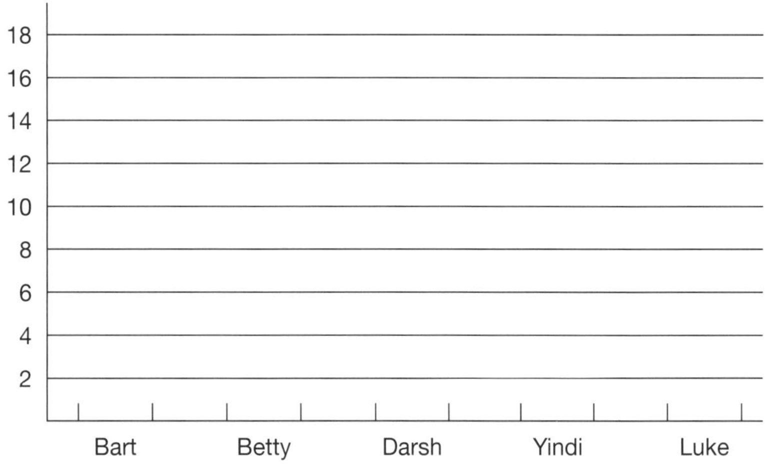

UNIT 9

Number and Algebra

SET 1 Basic

1 6×4

2 $21 \div 3$

3 6×8

4 $48 \div 6$

5 4×10

6 $\square \times 5 = 25$

7 $6 + 6 + 6$

8 $(9 + 3) - 4$

9 $(7 + 7) - 3$

10 $(2 \times 4) \times 3$

11 What is the product of 6 and 10?

12 What is the sum of 5, 4 and 8?

13 $24 \div 6$

14 $18 \div 9$

15 

There are 29 children in my class. If 17 are away sick, how many are well?

☐ children

SET 2 Division strategies

1

÷	360	120	480	240	600	300	540
6							

Use the halve and halve again strategy to divide by 4.

	Question	Halve	Halve
2	$40 \div 4$	20	10
3	$24 \div 4$		
4	$64 \div 4$		
5	$200 \div 4$		
6	$88 \div 4$		
7	$160 \div 4$		

Use the halve, halve again and halve again strategy to divide by 8.

	Question	Halve	Halve	Halve
8	$56 \div 8$	28	14	7
9	$80 \div 8$			
10	$200 \div 8$			
11	$160 \div 8$			
12	$96 \div 8$			

Space Three-dimensional descriptions

Colour the description and its matching object the same colour.

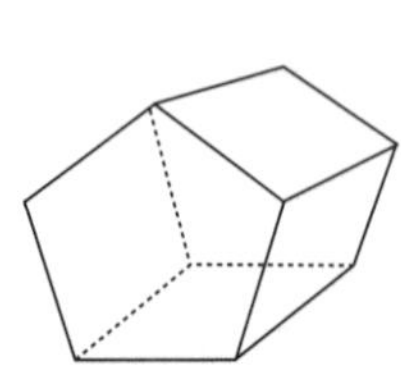

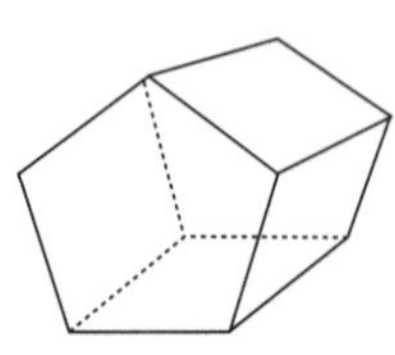

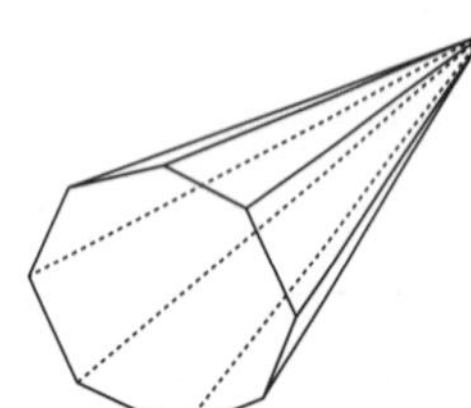

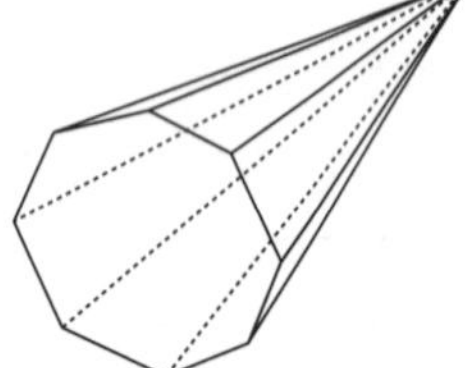

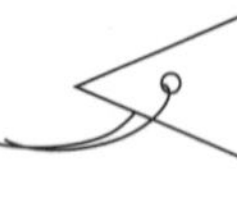

I am constructed from 2 pentagons and 5 rectangles. I am a prism.

I am constructed from 4 rectangles and 2 squares. I am a prism.

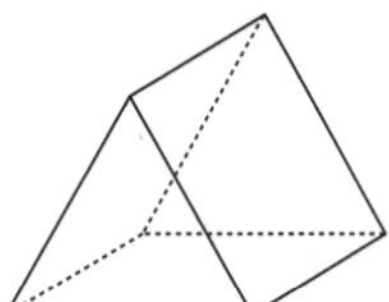

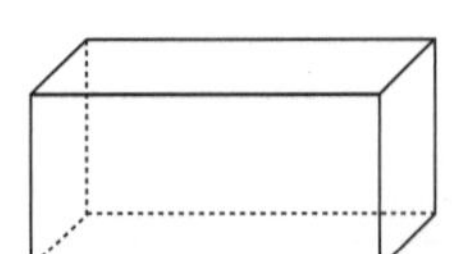

I am constructed from 2 equilateral triangles and 3 rectangles. I am a prism.

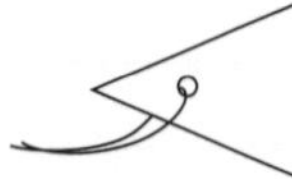

I am constructed from an octagon and 8 isosceles triangles. I am a pyramid.

Number and Algebra

SET 3 Mixed numerals

Draw a line to match the improper fractions to the mixed numbers.

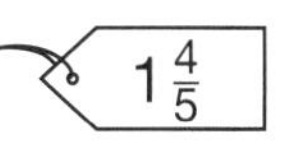

1	$\frac{6}{5}$	$1\frac{4}{5}$
2	$\frac{12}{10}$	$1\frac{1}{5}$
3	$\frac{14}{8}$	$1\frac{2}{10}$
4	$\frac{9}{5}$	$1\frac{6}{8}$
5	$\frac{17}{10}$	$1\frac{7}{10}$
6	$\frac{11}{8}$	$1\frac{3}{6}$
7	$\frac{9}{6}$	$1\frac{3}{8}$

Add these fractions to give more than 1 whole. Then convert the answers to mixed numbers.

8 $\frac{6}{8} + \frac{5}{8} = \frac{11}{8} =$

9 $\frac{4}{5} + \frac{3}{5} = \quad =$

10 $\frac{7}{10} + \frac{9}{10} = \quad =$

11 $\frac{4}{6} + \frac{4}{6} = \quad =$

12 $\frac{3}{8} + \frac{7}{8} = \quad =$

13 $\frac{8}{12} + \frac{10}{12} = \quad =$

True or false?

14 $1\frac{4}{5} < \frac{8}{5}$

15 $1\frac{6}{10} > 2$

16 $1\frac{3}{8} = \frac{13}{8}$

17 $1\frac{4}{6} > \frac{10}{6}$

SET 4 Extension

1 Estimate an answer to 287 + 409.

2 12.5 + 1.3

3 ☐ kg = 0.75 tonne

4 How much is 5 L at $4.50 per litre?

5 940 m + 59 m

6 (48 ÷ 6) + 463

7 $\frac{3}{8}$ of $960

8 A cricketer scored 4, 19, 27, 6. What was her average?

9 1302, 1332, 1362, ☐, ☐

10 How much is 500 g at $37 per kilogram?

11 How much change did I receive from $20 if I spent $1.65?

12 $107 + $52 + $218

13 $79.05 = ☐ c

Working Mathematically

14 Four girls have $60 between them. How much has each girl if:

Anna has twice as much as Biyu?	$
Biyu has 3 times as much as Deni?	$
Cath has $\frac{1}{6}$ of the total amount?	$
Deni has $\frac{1}{12}$ of the total amount?	$

Statistics and Probability Column graphs

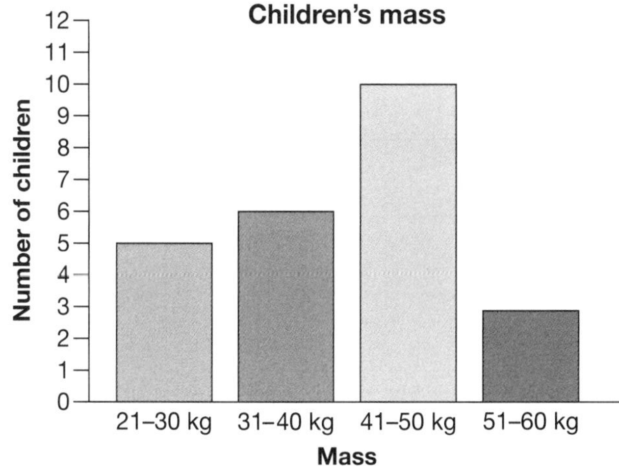

1 How many children had a mass between 31 and 40 kg?

2 How many children had a mass of 21 to 30 kg?

3 How many children had a mass of 51 to 60 kg?

4 How many more children had a mass of 41 to 50 kg than 21 to 30 kg?

5 Which category of mass did most children fit into?

Number and Algebra

SET 1 Basic

1 18 + 3

2 21 – 3

3 6 × 9

4 18 ÷ 3

5 20 ☐ 5 = 4

6 15 ☐ 3 = 5

7 20 ☐ 6 = 26

8 What is the product of 9 and 12?

9 22 ☐ 8 = 14

10 What is the sum of 18 and 10?

11 How many minutes in 2 hours?

12 How much are 3 lollies at 5c each?

13 $\frac{1}{2}$ of 100

14 Divide 12 by 12.

15

Each milk crate holds 16 L of milk. How many litres would 6 milk crates hold?

☐ L

SET 2 Subtraction

1

	TH	H	T	O
	5	8	3	4
–	4	2	7	3

2

	TH	H	T	O
	6	8	4	5
–	3	2	7	7

3

	TH	H	T	O
	2	5	0	8
–		4	2	8

4

	TH	H	T	O
	5	6	6	0
–		2	4	5

5 Subtract 137 from 899.

6 767 take away 347

7 907 minus 603

8 $9.75 – $3.30

9 Mr Bayu, the baker, baked 1050 cakes for the fete but sold only 742. How many were left? 

10 Jim needed to save $598 to buy a surfboard. If he saved $246, how much more did he need to save? 

Measurement Perimeter and area

Working Mathematically

Create any two shapes with an area of 6 cm^2. One should have a perimeter of 12 cm and the other a perimeter of 14 cm.

Number and Algebra

SET 3 Prime and composite numbers

Write prime or composite after each number.

1 25 ____________ 6 35 ____________

2 13 ____________ 7 40 ____________

3 32 ____________ 8 27 ____________

4 17 ____________ 9 31 ____________

5 42 ____________ 10 43 ____________

11 What is the next birthday you have coming up that is a prime number?

Working Mathematically

12 Can squared numbers be prime?

13 If you multiply two prime numbers, will the answer be a prime number?

14 What is the only prime number between 90 and 100?

15 Apart from 2, explain why even numbers can't be prime numbers.

____

____

SET 4 Extension

1 0.2 = 2 hundredths. True or false?

2 Write 29 in words.

3 How many $\frac{1}{4}$s in $2\frac{1}{4}$?

4 Write the factors for 45.

5 How many 25 cm lengths can be cut from 2 m?

6 How many surfaces has a pentagonal prism?

7 27 × 100

8 Are 18 and 50 multiples of 8?

9 How much is $2\frac{1}{2}$ kg of bacon at $6.40 per kilogram?

10 Average 10, 14 and 18

11 What is the sum of multiples of 3 between 11 and 16?

12 Write these numbers in descending order: 27 301, 28 963, 35 427.

13 Which is larger: $\frac{3}{4}$ or 0.5?

14 How much are 9 apples at 25c each?

15 Write 20 206 in words.

Statistics and Probability Likelihood

36 pieces of fruit were placed in a grocery bag. Use the data shown on the Fruit Wheel to decide whether these statements about the type of fruit to be selected first from the bag are true or false.

		True or false
1	Orange and strawberry have an even chance.	
2	Apples are certain to be selected first.	
3	Mangoes are less likely than pears.	
4	You are more likely to pick an orange than an apple.	
5	It is impossible to select a plum.	

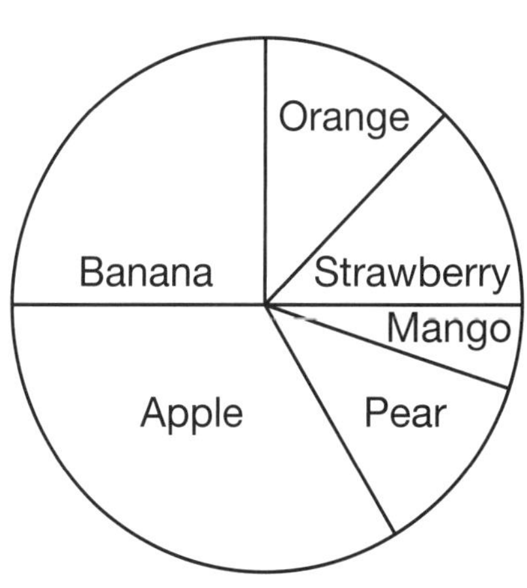

UNIT 11

Number and Algebra

SET 1 Basic

1 17 + 5

2 26 ☐ 4 = 30

3 30 – 15

4 24 ☐ 8 = 16

5 7×5

6 26 ☐ 2 = 13

7 $24 \div 3$

8 9 ☐ 3 = 27

9 195, 200, 205, ☐

10 1 minute = ☐ seconds

11 What is the product of 9 and 7?

12 What is the difference between 23 and 7?

13 How much are 7 lollies at 8c each?

14 How many years in 1 century?

15

Zorba had 7 boxes of chocolates with 30 chocolates in each. How many chocolates did Zorba have?

☐ chocolates

SET 2 Multiplication and division

Write a multiplication fact to check each division.

1 $21 \div 3 = 7$ ✓ $7 \times 3 = 21$

2 $49 \div 7 = 7$ ☐ ______

3 $54 \div 6 = 9$ ☐ ______

4 $54 \div 8 = 7$ ☐ ______

5 $36 \div 4 = 8$ ☐ ______

6 $81 \div 9 = 9$ ☐ ______

7 $56 \div 8 = 6$ ☐ ______

8 $45 \div 5 = 9$ ☐ ______

9 $42 \div 6 = 7$ ☐ ______

10 $3\overline{)72}$

11 $6\overline{)96}$

12 $7\overline{)79}$

13 $2\overline{)54}$

14 $5\overline{)83}$

15 $5\overline{)90}$

16 $3\overline{)78}$

17 $7\overline{)88}$

18 $4\overline{)66}$

Number and Algebra Expanding numbers

Write the number that is equal to:

1 5 tens ______

2 10 tens ______

3 15 tens ______

4 5 hundreds ______

5 10 hundreds ______

6 15 hundreds ______

7 5 thousands ______

8 10 thousands ______

9 15 thousands ______

10 Use the numeral expander to record the number **59632**.

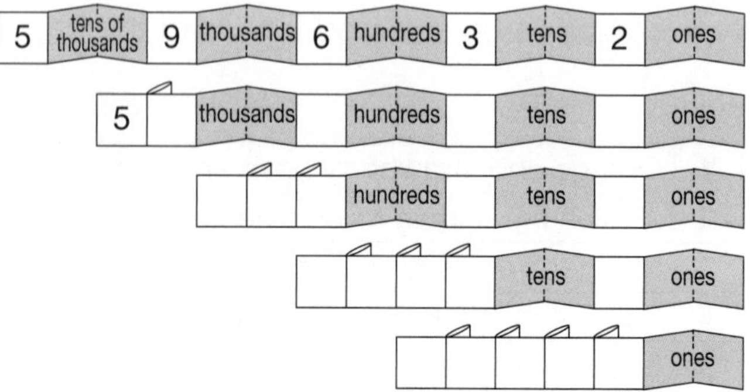

Number and Algebra

SET 3 Decimals

Record the decimals.

		Decimal		
		Ones	Tenths	Hundredths
1		0 .		
2		.		
3		.		

Shade the grids to match the decimals.

4 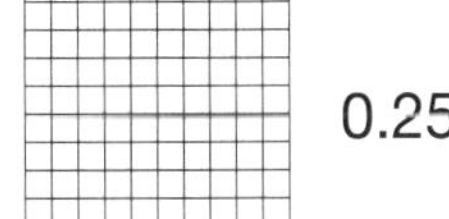 0.25

5 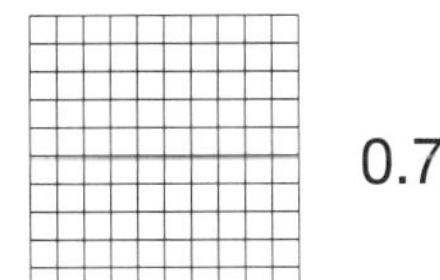 0.79

6 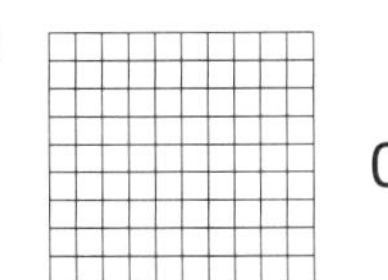 0.10

7 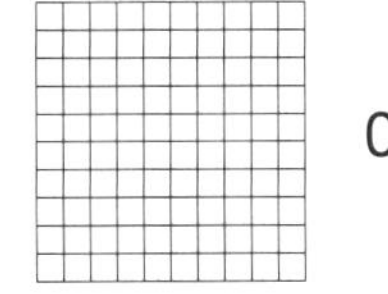 0.80

SET 4 Extension

1 $4\frac{1}{4}$ m = ☐ centimetres

2 Are 27 and 45 multiples of 9?

3 How many 750 g packets make $4\frac{1}{2}$ kg?

4 $\frac{7}{8}$ of 168

5 Write the numeral for thirty-nine.

6 How much are 5 boxes of chocolate at $2.20 each?

7 37×8

8 How much is 4 kg of steak at $2 per $\frac{1}{2}$ kilogram?

9 How many 60 cm lengths in 3 m?

10 Write these numbers in ascending order: 53 224, 20 206, 7496

11 What is the perimeter of a triangle with three 14 cm sides?

12 How many hundreds in 4230?

Working Mathematically

60 km/h

13 What is the maximum distance a driver could go in $2\frac{1}{2}$ hours if they kept to the speed limit?

Measurement Square centimetres

Calculate the area of these shapes.

1 

Area = ☐ cm^2

2 

Area = ☐ cm^2

3 

Area = ☐ cm^2

Number and Algebra

SET 1 Basic

1 4 × 11

2 \$3 × 4

3 68 = ☐ tens + ☐ ones

4 \$4 + \$8 + \$5

5 5^2

6 200 + 30 + 9

7 \$9.70 – \$2

8 7^2

9 How much are 2 brushes at \$1.30 each?

10 4 × 1000

11 Divide 27 by 3.

12 Multiply 6 by 5.

13 \$4 × 10

14 Add 40, 10 and 50.

15 Sarah's step is 75 cm. How many metres will she cover in 10 steps? ☐ m

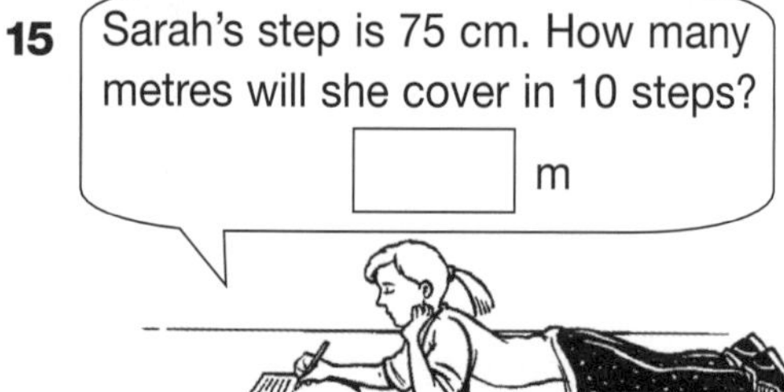

SET 2 5-digit addition

\$35 500	\$19 999	\$42 600
Mini-bus	2-door car	Truck

\$25 444	\$21 425	\$29 999
Sports car	Taxi	Mini-van

Calculate the value of each pair of vehicles.

1

Taxi	\$
Sports car	\$
	\$

2

Truck	\$
2-door car	\$
	\$

3

Mini-bus	\$
Sports car	\$
	\$

4

2-door car	\$
Mini-van	\$
	\$

Statistics and Probability Line graphs

Ted's walking graph

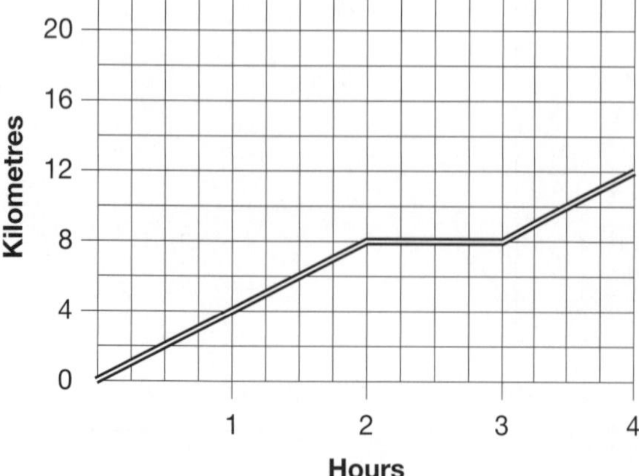

1 How far had Ted walked after 1 hour?

2 How far had Ted walked after 3 hours?

3 How far had Ted walked after 4 hours?

4 How far did Ted walk between the 2nd and 3rd hour?

5 What was Ted's average speed for the walk?

Number and Algebra

SET 3 Rounding money

- Prices ending in 1 cent and 2 cents are rounded down to 0 cents.
- Prices ending in 3 cents and 4 cents are rounded up to 5 cents.
- Prices ending in 8 cents and 9 cents are rounded up to 10 cents.
- Other prices are rounded up or down to the nearest 5 cents, depending on the amount in the ones column.

Round these amounts to the closest 5 cents.

1 $2.24

2 $3.61

3 $6.88

4 $9.62

5 $8.37

6 $11.09

7 $15.02

8 $21.01

9 $49.99

10 

$4.02

Is $4 enough to pay for the juice?

____

SET 4 Extension

1 ☐ ÷ 8 = 7

2 $2.30 + $1.25 + $1.48

3 $(\frac{3}{5} \times 20) + 250$

4 What is 57 more than 1374?

5 What is the perimeter of a square with 17 cm sides?

6 How much change did I receive from $20 if I spent $4.78?

7 Round 6538 to the nearest thousand.

8 If 7 cost $2.80, how much for 9?

9 Share $4.80 among 3 people.

10 Write the multiples of 7 between 13 and 50.

11 How much are 6 pens at $7.10 each?

12 How many tenths in $7\frac{1}{2}$?

13 How many minutes from 9:55 to 12:30?

14 $(8 \times 10^2) + (6 \times 10)$

15 How many hundreds in 15 287?

16 How many 15 mL medicine bottles can be filled by a 300 mL jug?

17 Round to the nearest whole km to answer 15.15 km + 27.89 km.

Measurement The cubic centimetre

How many cubic centimetres (centicubes) would be needed to fill each box?

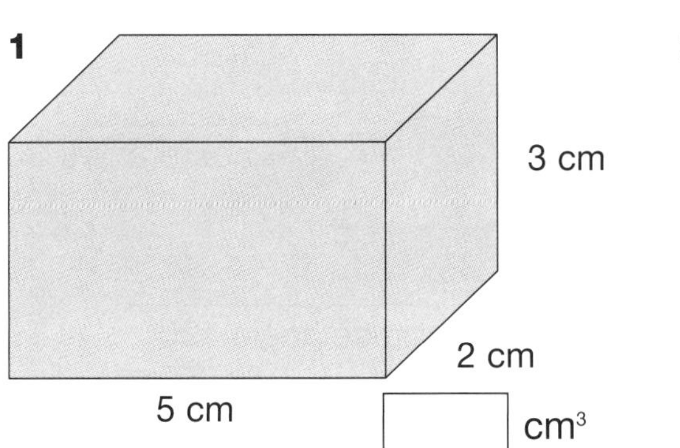

2

3 cm

3 cm

8 cm ☐ cm³

UNIT 13

Number and Algebra

SET 1 Basic

1 3, 6, 9, ☐, 15

2 10×9

3 8×5

4 $16 \div 4$

5 $28 \div 4$

6 $(2 + 5) \times 3$

7 $(3 + 3) \times 4$

8 $600 + 8$

9 $100 - 10$

10 $5^2 - 2$

11 How much are 7 pens at $1.10 each?

12 What is one-quarter of 40?

13 $6.21 = ☐ c

14 $10 × 3

15 Mark planted 7 rows of flowers with 8 in each row. How many flowers did he plant?

☐ flowers

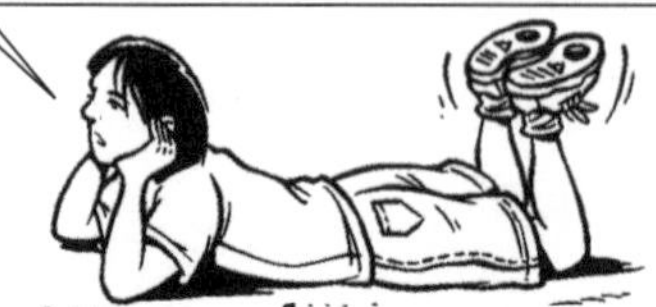

SET 2 Multiplication

1

	TH	H	T	O
		3	6	4
×				3

2

	TH	H	T	O
		4	9	6
×				5

3

	TH	H	T	O
		2	8	7
×				7

4

	TH	H	T	O
		6	0	9
×				5

5

	TH	H	T	O
		3	8	5
×				6

6

	TH	H	T	O
		3	4	5
×				8

7 Seon had 8 boxes with 137 tennis balls in each. How many tennis balls did she have altogether?

8 Deshan collected an average of 98 stamps each month for 9 months. If they are worth an average of $2 each, what is the value of his stamps?

Space Measuring angles

1 Measure these angles.

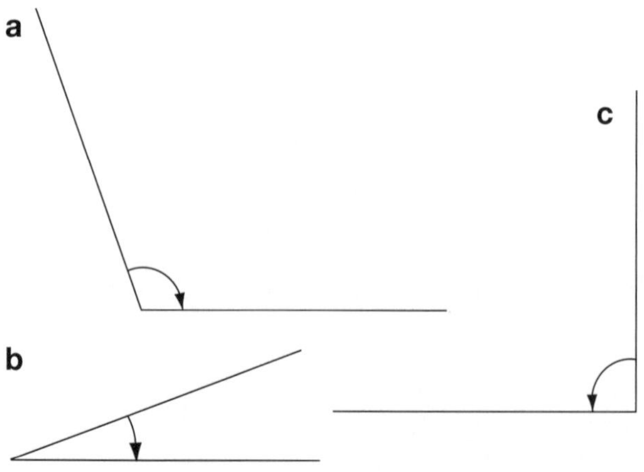

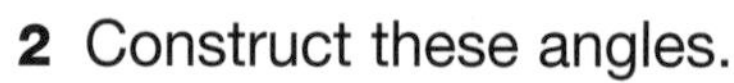

2 Construct these angles.

a acute angle 60°

b obtuse angle 120°

c acute angle 30°

Number and Algebra

SET 3 Missing numbers

Find the missing numbers.

1 12 + ☐ = 36

2 50 − ☐ = 27

3 3 × 7 = 31 − ☐

4 6 × ☐ = 66

5 60 − 25 = 7 × ☐

6 54 ÷ ☐ = 9

7 4 × ☐ = 16 + 16

8 23 + 17 = 4 × ☐

9 (30 ÷ ☐) × 4 = 12

10 (2 + ☐) + 13 = 20

Draw a line from the number sentence to the missing number.

11 38 − ☐ = 20	9
12 7 × ☐ = 58 − 9	18
13 16 + 20 = 4 × ☐	7
14 (24 ÷ 2) + ☐ = 27	15
15 (5 + ☐) ÷ 8 = 2	3
16 32 + (21 ÷ ☐) = 39	11

SET 4 Extension

1 25 × 1000

2 What is the sum of 1323 and 506?

3 List the factors of 42.

4 $(\frac{7}{8} \times 64) + (\frac{3}{4} \times 120)$

5 What is the difference between 357 and 203?

6 How much is $3\frac{1}{4}$ kg of meat at $12 per kilogram?

7 What is the value of 7 in 2.74?

8 Round 38 126 to the nearest thousand.

9 Which is larger: 7^2 or 10 × 9?

10 4000 + 600 + 50 + 8

11 How much are 6 apples at 3 for $1.25?

12 6:25 pm + 45 minutes

13 If a square has a perimeter of 24 cm, what is the length of each side?

Working Mathematically

14 Tina bought 10 packets of chocolate biscuits. Can you calculate the cost faster than using a calculator?

Biscuits	$3.95
Biscuits	$3.95
Biscuits	$3.95
Biscuits	$3.95
Biscuits	$3.95
Biscuits	$3.95
Biscuits	$3.95
Biscuits	$3.95
Biscuits	$3.95
Biscuits	$3.95

Measurement 24-hour time

Write the starting time of each program.

1 *The Last Frontier* ☐

2 *The Midday Show* ☐

3 *Days of Our Lives* ☐

4 *Who's the Boss?* ☐

5 *A Current Affair* ☐

6 If a TV was set to record a program at 19:30, what program would be recorded?

7 If the TV was set to record at 22:30, what program would be recorded?

6:00	**Tennis** – The US Open.
9:30	**The Last Frontier** – Wildlife.
10:00	**Romper Room** – For kids.
10:30	**Police Story** – (PGR, Final).
11:30	**Wheel of Fortune** – Game.
12:00	**The Midday Show** – Topical.
1:30	**Days of Our Lives** (PGR).
2:35	**The Young And The Restless.**
3:30	**Focus on Living** – Religious.
4:00	**Who's The Boss?** – (Return).
4:30	**Bush Beat** – For children.
5:00	**Happy Days** – Comedy series.
5:30	**Sale Of The Century** – (*S).
6:00	**News, Sport And Weather.**
7:00	**A Current Affair** – Topical.
7:30	**A Country Practice** – (*S, PGR).
8:30	**Twin Peaks** – (PGR). The identity of Laura Palmer's killer is revealed; Leland tries to discourage Maddy from going home.
9:30	**MacGyver** – (PGR). Drama.
10:30	**Night Court** – (PGR). Comedy.
11:00	**Motorcycle Racing** – Le Mans 500cc Grand Prix.
12:00	**Close.**

UNIT 14

Number and Algebra

SET 1 Basic

1 5 × 8

2 42 ☐ 7 = 6

3 9 ☐ 9 = 81

4 465, 455, ☐, 435

5 6 chocolate bars at 60c each

6 One half of 90

7 36 ÷ 9

8 Multiply 7 by 11.

9 What is the sum of 60 and 25?

10 How many eggs in 2 dozen?

11 $10^2 + 7$

12 What is the difference between 53 and 39?

13 What is the value of 2 in 60 208?

14 (6 + 3) × 5

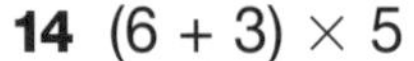

15 

☐

SET 2 Subtraction strategies

Subtract the thousands first, then the hundreds followed by the tens and ones.

9548 – 3426
Subtract 3000 = 6548
Subtract 400 = 6148
Subtract 20 = 6128
Subtract 6 = 6122

1 8959 – 6737

2 8348 – 7216

3 7433 – 5320

4 6528 – 4416

5 9486 – 7265

$1510

$1025

Calculate how much money Lee has left after buying each item if she began the day with $4999.

	Purchase	Money left
6	Lounge	
7	TV	
8	Computer	

Space Describing objects

Write a description of each object.

____

____

____

____

____

____

Number and Algebra

SET 3 Patterns and rules

Apply the rules to complete the patterns.

1 Rule: × 3 + 1

2	4	6	8	10	12	14	16

2 Rule: × 5 – 8

10	11	12	13	14	15	16	17

3 Rule: × 4 – 2

2	4	6	8	10	12	14	16

4 Rule: ÷ 5 + 2

95	90	85	80	75	70	65	60

5 Work backwards to find the input numbers.

Rule: × 4 – 3

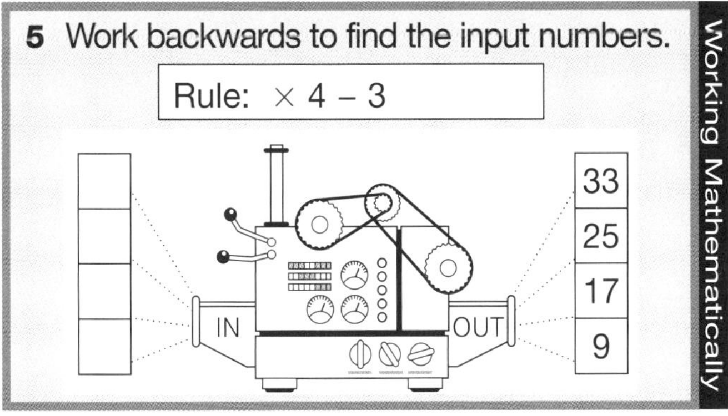

Working Mathematically

SET 4 Extension

1 Round 18 627 to the nearest 1000.

2 ☐ ÷ 9 = 11

3 \$6.50 × 100

4 List the factors of 32.

5 How many hours from 6 am to 3:30 pm?

6 How much change do I receive from \$25 if I spent \$22.45?

7 How many months in $3\frac{1}{3}$ years?

8 Which is larger: $\frac{1}{2}$, 80% or $\frac{60}{100}$?

9 If 5 cost \$2.00, how much do 2 cost?

10 If a square has an area of 36 cm^2, what is the length of the base?

11 How many 400 g bags in 6 kg?

12 6×5^2

13 Share \$6.30 among 7 people.

14 How many 20 cm lengths in 1.6 m?

15 It took 5 minutes to run one lap. How many laps could be run in 2 hours?

16 Give the dimensions in metres of a square whose perimeter and area are the same number.

Working Mathematically

Measurement Perimeter

Measure the following perimeters in centimetres.

1

cm

2

cm

3

cm

UNIT 15

Number and Algebra

SET 1 Basic

1 1290, 1295, 1300, ☐

2 27 – 6

3 35 ☐ 7 = 42

4 18 + 9

5 50 ☐ 30 = 20

6 6 × 9

7 48 ÷ 8

8 6 ☐ 7 = 42

9 18 ☐ 6 = 3

10 What is the product of 7 and 8?

11 How many millilitres in 1 L?

12 5 pencils at 5c each

13 What is the value of 7 in 37 256?

14 How many years in a decade?

15 

Wayan picks 23 coconuts an hour. How many will he pick in 7 hours?

☐ coconuts

SET 2 Expanding numbers

1 Write the number for twenty-six thousand, two hundred and seven.

2 Write the number for eighty thousand and sixty-seven.

3 Write the number ten more than 754 165.

4 Write the number 20 more than 63 347.

5 Add 10 000 to 333 500.

6 Write 26 007 in words. ______________________

____

7 Write 510 150 in words. ______________________

____

8 Which is greater: 9000 or 9 ten thousands?

9 Round 63 795 to the nearest thousand.

10 Round 63 795 to the nearest 10 000.

11 What is 3000 more than 760 518?

12 301 507 = ☐ + ☐ + ☐ + ☐

Statistics and Probability Databases

Netball database

	First name	Surname	Address	Age	Grade
1	Jill	Arnott	Shellharbour	11	11A
2	Amal	Brown	Shellharbour	10	10B
3	Kelly	Cross	Kiama	14	14C
4	Lauren	Ellis	Dapto	12	12A
5	Kate	Hood	Figtree	11	11A
6	Sophie	Jones	Warilla	11	11B
7	Eva	Kelly	Figtree	15	15C
8	Lena	Piper	Dapto	12	12A
9	Soula	Rhodes	Shellharbour	11	11A
10	Kimberley	Smith	Figtree	16	16A
11	Auril	Suvan	Warilla	14	14B
12	Tim	Zatt	Figtree	10	10B

1 How many children live in Shellharbour?

2 How many children are 11 years old?

3 How many children live in Dapto?

4 How many children play in the 11A grade?

5 How has this database been organised?

Number and Algebra

SET 3 Fractions on a number line

Complete the number lines.

1

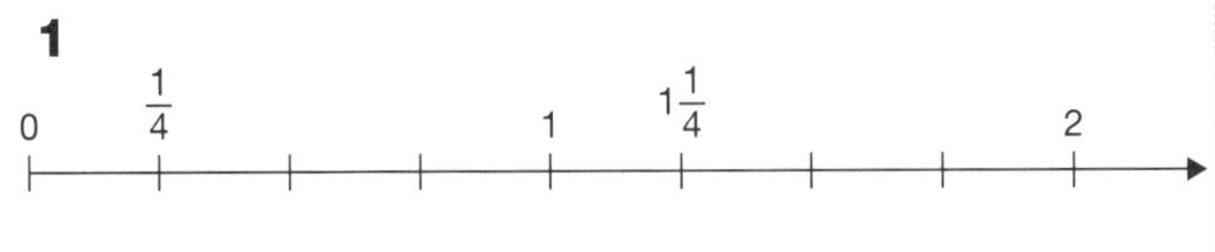

2

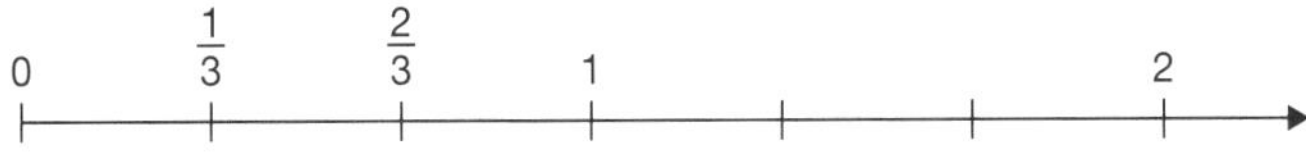

3

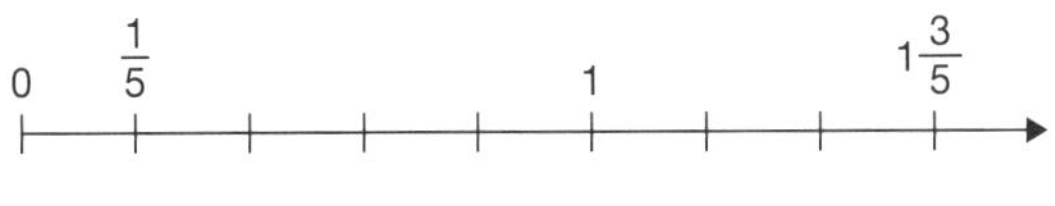

4

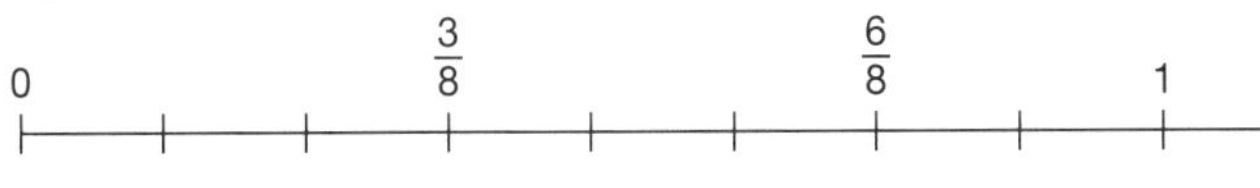

5

0 $\frac{1}{10}$ $\frac{2}{10}$

6

0 $\frac{1}{2}$ 1 $1\frac{1}{2}$

SET 4 Extension

1. $\frac{3}{10}$ of 200
2. Round 21 350 to the nearest 1000.
3. 9999 + 801
4. $8\frac{1}{2}$ km = ______ m
5. Write the numeral for twenty-seven thousand, two hundred and twenty-six.
6. 96×70
7. List prime numbers between 20 and 30.
8. Which is larger, 2.5 or $2\frac{1}{10}$?
9. How many minutes from 3:25 to 7:00?
10. Which is larger, $2\frac{1}{3}$ or $2\frac{1}{4}$?
11. What is the value of 9 in 3.79?
12. What is the perimeter of a rectangle with length 7 cm and width 3 cm?
13. How many 35 cm lengths in 7 m?
14. Average of 15, 30, 45, 70
15. How much is 500 g of meat at $7 per kilogram?
16. Alexis runs 2 km in 10 minutes. How far will she run in 65 minutes?

Measurement Area in square metres

Calculate the areas of these shapes, using the scale.

SCALE 1 cm = 1 m

1

Blackboard ___ m^2

2

Lounge room ___ m^2

3

Swimming pool ___ m^2

UNIT 16

Number and Algebra

SET 1 Basic

1 4, 8, ☐, ☐, 20

2 9×7

3 8×6

4 $20 \div 5$

5 $30 \div 6$

6 $(3 + 4) \times 3$

7 $(4 + 2) \times 5$

8 $600 + 70 + 3$

9 $200 - 30$

10 Subtract 20 from 80.

11 Subtract 3 tens from 40.

12 What is the sum of 150 and 46?

13 What is the product of 200 and 10?

14 How much are 3 magazines at $1.20 each?

15 

How much are 3 bottles of soft drink at $2.20 a bottle?

$ ☐

SET 2 Division

1 $2\overline{)864}$

2 $4\overline{)488}$

3 $6\overline{)964}$

4 $4\overline{)886}$

5 $3\overline{)690}$

6 $4\overline{)840}$

7 $3\overline{)972}$

8 $6\overline{)684}$

9 $6\overline{)786}$

10 $4\overline{)968}$

11 $3\overline{)762}$

12 $7\overline{)868}$

13 Hanny won $756 in the lotto, which he had to share with another 4 people. How much did each person receive? 

14 George saved $459 in 3 months. What was his average saving per month? 

Space Measuring angles

1 Which angles measure 90°? ____________

2 Which angle measures 45°? ____________

3 Which angles measure 110°? ____________

4 Which angle measures 180°? ____________

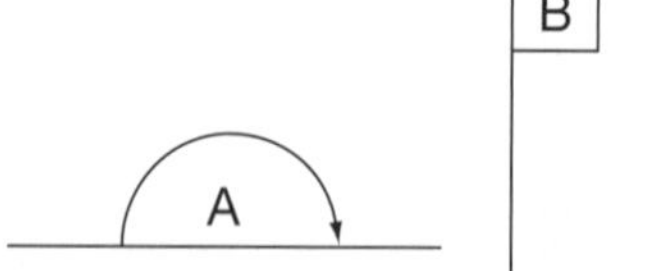

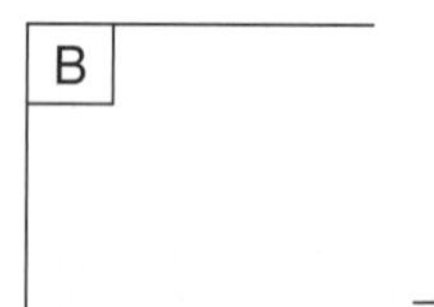

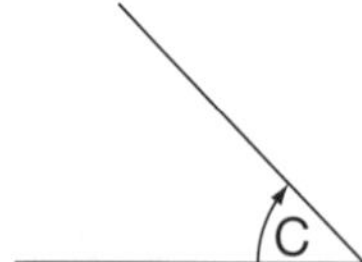

True or false?

5 Obtuse angles are between 90° and 180°. ____________

6 Acute angles are less than 90°. ____________

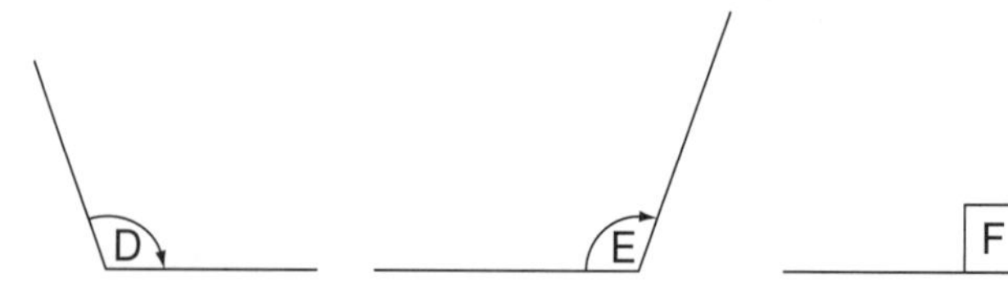

Number and Algebra

SET 3 Factors

Find four factors for each number.

	Number	Factors
1	20	
2	16	
3	12	
4	40	
5	32	
6	30	

7 Is 3 a factor of 12?

8 Is 8 a factor of 32?

9 Is 6 a factor of 27?

10 Is 6 a factor of 18?

11 Find all the factors for 24.

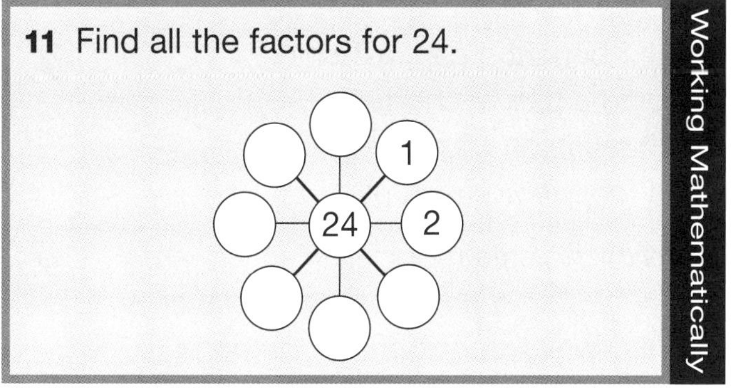

Working Mathematically

SET 4 Extension

1 Share 40 among 6 people.

2 $518 × 100

3 Double $13.49.

4 What is the area of a square with 5 cm sides?

5 What is the difference between 408 and 162?

6 $8.78 × 8

7 $4\frac{1}{4}$ m = ☐ cm

8 50% of 80

9 Which is larger: 7^2 or $(7 \times 6) + 7$?

10 Write $\frac{7}{100}$ as a decimal.

11 27.6 – 4.3

12 If 3 cost 45c, how much would 7 cost?

13 If a square has a perimeter of 36 cm, what is the length of the base?

14 Write thirty-seven thousand, nine hundred and forty-two in figures.

15 A palindromic number reads the same from right to left as it does from left to right, e.g., 3 4 5 4 3.

Make a palindromic number from these numbers: 2, 6, 3, 3, 2.

Working Mathematically

Space Symmetry and rotational symmetry

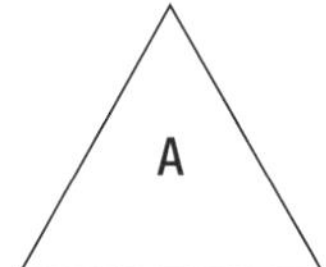

B

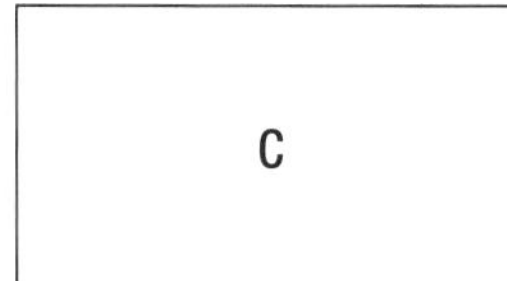

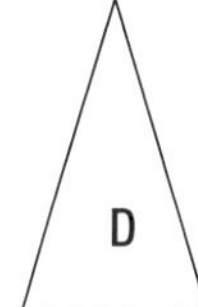

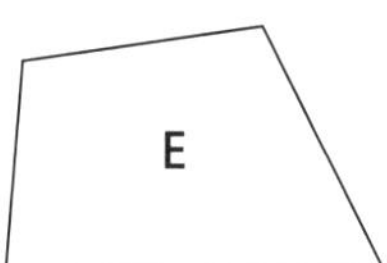

1 Which shapes above have rotational symmetry? ____________

2 Which shape has 2 lines of symmetry? ____________

3 Which shape has 1 line of symmetry? ____________

Number and Algebra

SET 1 Basic

1 ☐ × 5 = 25

2 32 ÷ 8

3 20 ☐ 8 = 28

4 28 – 16

5 $6^2 + 2$

6 (3 × 4) + 2

7 7 ☐ 5 = 35

8 (2 × 6) + 4

9 Divide 28 by 7.

10 $7.85 = ☐ c

11 115, 119, 123, ☐

12 What is the difference between 75 and 44?

13 Divide 12 by 12.

14 How much are 2 magazines at $2.10 each?

15

SET 2 Addition problems

$11.65

$7.24

$29.20

$32.40

1 How much would it cost to buy the complete outfit?

2 How much would it cost to buy 2 pairs of socks and a T-shirt?

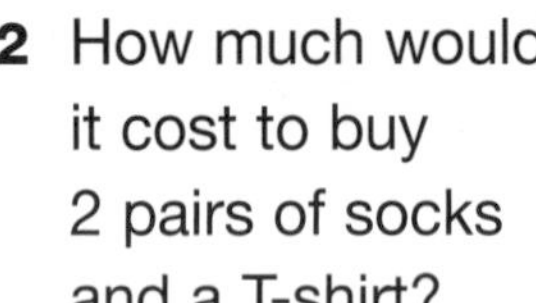

3 Petra spent $36.44 on 2 items. What did she buy?

4 If I bought a cap, a T-shirt and a pair of shorts, how much change would I receive from $100?

5 Which costs more:

a a T-shirt and a pair of socks or

b two caps and a pair of shorts?

Statistics and Probability Chance

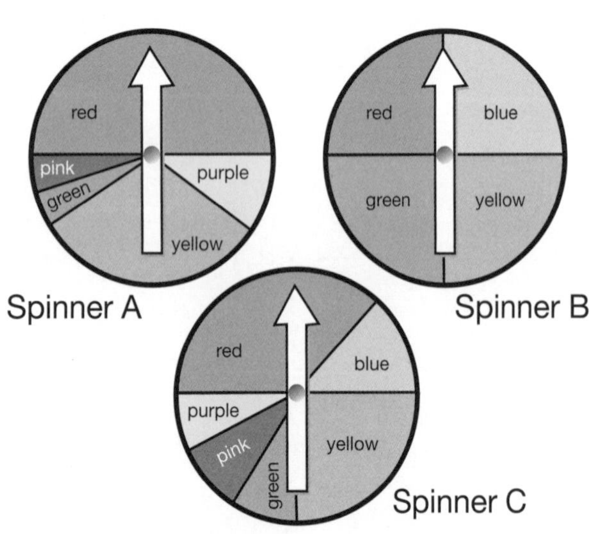

1 Which spinner has the most chance of landing on red?

2 Which spinner has the least chance of landing on green?

3 Which spinner has an even chance of landing on red, blue, yellow or green?

4 Which spinner has an even chance of landing on purple, pink or green?

5 Which spinners have the same chance of landing on yellow?

Number and Algebra

SET 3 Adding and subtracting fractions

Complete the calculations.

1 $\frac{1}{4} + \frac{1}{4}$

2 $\frac{8}{10} - \frac{5}{10}$

3 $\frac{1}{3} + \frac{1}{3}$

4 $\frac{11}{12} - \frac{4}{12}$

5 $\frac{3}{5} + \frac{1}{5}$

6 $\frac{9}{10} - \frac{3}{10}$

7 $\frac{3}{10} + \frac{6}{10}$

8 $\frac{45}{100} - \frac{31}{100}$

9 $1\frac{1}{5} + \frac{1}{5}$

10 $2\frac{3}{5} - \frac{2}{5}$

11 $3\frac{4}{10} + \frac{3}{10}$

12 $4\frac{9}{10} - \frac{5}{10}$

13 $5\frac{3}{8} + \frac{2}{8}$

SET 4 Extension

1 (49 ÷ 7) + 509

2 Average 30, 35, 27 and 40

3 50% of 460

4 29 × 100

5 Write these numbers in descending order: 29 501, 48 602, 37 100.

6 What is the area of a rectangle with sides of 7 cm and 4 cm?

7 How many hundreds in 5280?

8 How many minutes from 7:45 am to 10:15 am?

9 20% of $45

10 $(6 \times 10^2) + (4 \times 8)$

11 Round 19 649 to the nearest thousand.

12 Share 97 between 6 people.

13 3.12, 3.18, 3.24, ☐, ☐

14 Find the lowest common multiple of 9 and 5.

15 Write the numeral for a quarter of a million.

☐

16 If a rectangle has a perimeter of 90 cm and one side is 17 cm, what is the length of the other sides that are not 17 cm?

Space Three-dimensional views

Colour the top views red, the side views blue and the front views green.

A

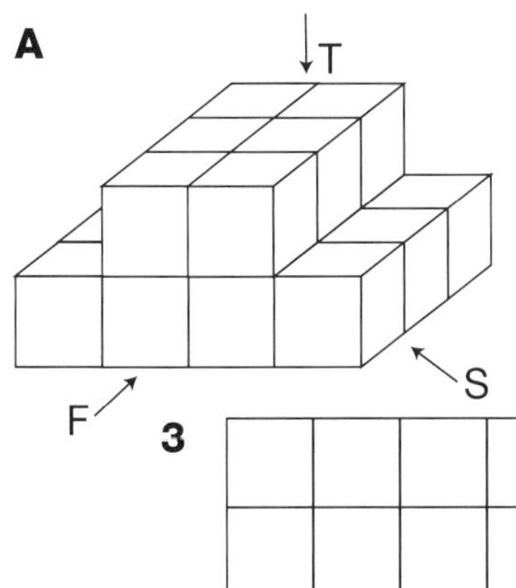

1

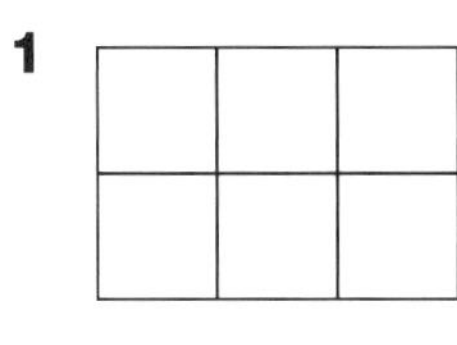

2

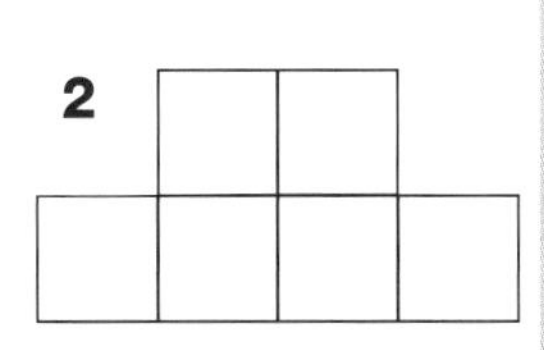

3

B

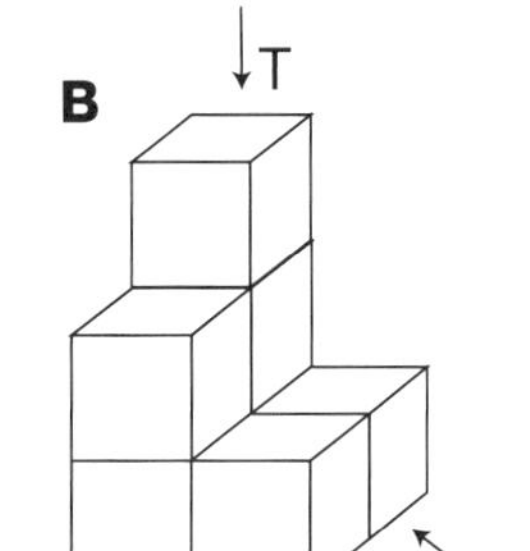

1

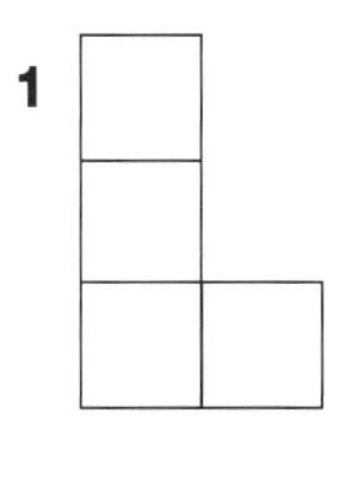

2

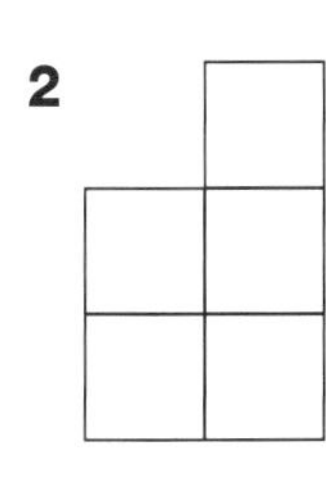

3

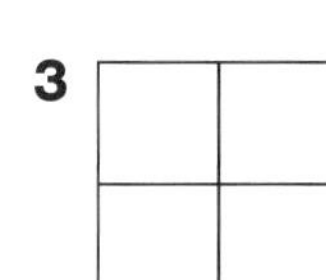

Number and Algebra

SET 1 Basic

1 6 × 9

2 11 − 5

3 14 + 3

4 35 ÷ 7

5 28 ☐ 4 = 7

6 56 ☐ 8 = 64

7 60 ☐ 12 = 48

8 7 ☐ 6 = 42

9 213, 219, 225, ☐

10 180 minutes = ☐ hours

11 What is the value of 3 in 31 246?

12 How much are 3 rulers at 25c each?

13 Divide 40 by 5.

14 What is the product of 9 and 6?

15 Elle has 12 orange quarters cut up for the netball team. How many oranges did she cut up?

 oranges

SET 2 4-digit subtraction

Crack the code.

3033	4745	2472	4835	2500	1187	1409	6545
L	R	S	E	O	P	A	W

1
```
  3 5 6 4
− 1 0 9 2
```

2
```
  6 5 9 0
− 4 0 9 0
```

3
```
  5 6 0 5
− 2 5 7 2
```

4
```
  8 1 2 5
− 6 7 1 6
```

5
```
  7 6 4 3
− 2 8 9 8
```

6
```
  5 0 0 2
− 3 8 1 5
```

7
```
  6 2 6 2
− 3 7 6 2
```

8
```
  7 2 8 0
−   7 3 5
```

9
```
  4 9 0 2
−     6 7
```

10
```
  9 7 8 4
− 5 0 3 9
```

1	2	3	4	5

6	7	8	9	10

Statistics and Probability Line graphs

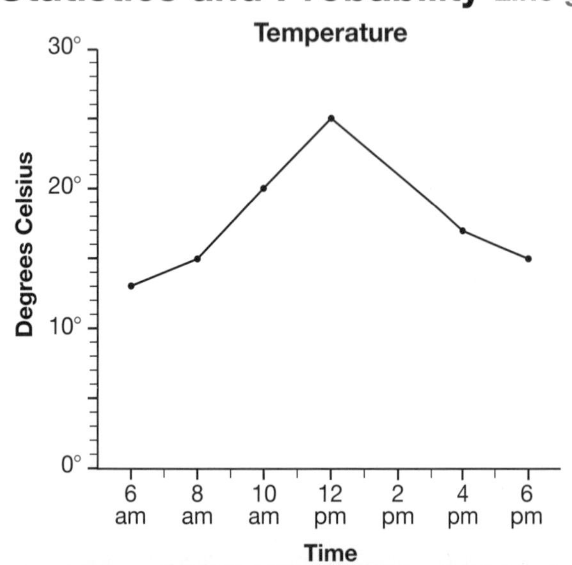

1 What was the temperature at 10 am? ☐ °C

2 What was the temperature at 4 pm? ☐ °C

3 How much did the temperature rise between 8 am and 12 pm? ☐ °C

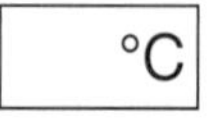

4 What is the difference between the highest and lowest temperatures? ☐ °C

5 Estimate the temperature at 2 pm. ☐ °C

Number and Algebra

SET 3 Order of operations

1 $3 + 6 \times 2 =$

2 $(3 + 6) \times 2 =$

3 $2 \times 7 + 5 =$

4 $2 \times (7 + 5) =$

5 $3 \times 5 \times 2 =$

6 $2 \times 3 \times 4 + 7 =$

7 $3 + 2 \times 7 + 3 =$

8 $30 - 3 \times 5 =$

9 $(10 - 7) \times 5 - 6 =$

10 $40 - 7 \times (8 - 6) =$

Write true or false.

11 $3 \times 7 + 9 = 30$ ________

12 $3 \times (7 + 9) = 30$ ________

13 $5 \times 8 + 16 = 56$ ________

14 $5 + 3 \times 9 + 7 = 39$ ________

15 $10 + 5 \times 6 + 13 = 77$ ________

16 $80 - 3 \times 10 + 7 = 57$ ________

17 $3 \times 9 - 10 - 8 = 10$ ________

18 $250 - 9 \times 10 - 70 = 190$ ________

19 $300 - 7 \times 9 + 63 = 237$ ________

20 $400 - 7 \times (5 + 6) = 223$ ________

Working Mathematically

21 (□ + △) × 5 = ⬡60

Supply your own numerals to complete this number sentence. You are not allowed to use a numeral more than once.

SET 4 Extension

1 $2^2 + 5^2$

2 79×1000

3 Write prime numbers between 6 and 20.

4 Write $4\frac{17}{100}$ as a decimal.

5 $\frac{4}{5}$ of 1 kg

6 How many minutes between 4:23 and 6:37?

7 Round 37 212 to the nearest hundred.

8 What is the perimeter of a hexagon with 27 cm sides?

9 25% of 60

10 Write 149 in words.

11 $0.5 = \frac{}{10} = \frac{}{100} = \frac{}{2}$

12 If 5 cost 75c, how much would 9 cost?

13 How many seconds in $2\frac{1}{4}$ minutes?

14 Write these numbers in descending order: 1479, 256, 35 256.

Working Mathematically

15 

If 1 mL of water has a mass of 1 g and volume of 1 cm^3, what would be the mass of a 2 L bottle?

Measurement Millilitres and cubic centimetres

How many times would each container need to be poured into the 2 L bucket to fill it?

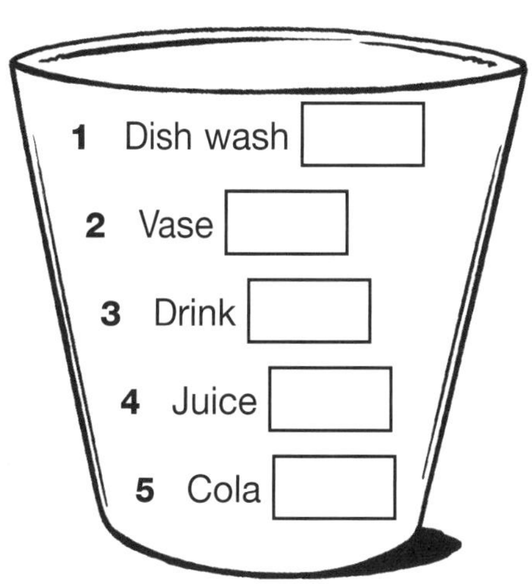

Number and Algebra

SET 1 Basic

1 14 + 6 + 5

2 19 – 5

3 6 × 8

4 40 ÷ 8

5 What is the product of 6 and 3?

6 What is the sum of 6, 4 and 9?

7 $4.15 = ☐ c

8 200 – 5

9 Double 12.

10 13 + 4 + 5

11 20 – 3

12 7 × 9

13 42 ÷ 2

14 What is the product of 7 and 4?

15 Renata had $8 and spent one-quarter of it. How much did she have left?
$ ☐

SET 2 Division

Number cross

1			2	3		4	5
		6		7			
	8					9	
10		11	12		13		
14							
		15					16

Across

1 84 ÷ 7

2 96 ÷ 4

4 76 ÷ 4

7 325 ÷ 5

8 426 ÷ 6

11 108 ÷ 9

13 920 ÷ 5

14 232 ÷ 8

15 882 ÷ 9

Down

1 51 ÷ 3

3 138 ÷ 3

5 376 ÷ 4

6 999 ÷ 9

9 861 ÷ 3

10 889 ÷ 7

12 832 ÷ 4

15 837 ÷ 9

16 518 ÷ 7

Statistics and Probability Ordinal data

The chart below shows the names of the children in the 5 group levels at Aqua Swim School.

Frogs	Seals	Dolphins	Sharks	Whales
Zara	Isla	Mara	Ava	Noah
Eve	Will	Leo	Joni	Vaya
Kirra	Theo	Waru	Usman	Grace
	Levi	Rumi	Rua	Zac
	Royston	Aran	Tane	
	Mali		Ollie	
	Jett		Jack	
	Koen		Sam	
			Lulu	
			Tui	

Complete the column graph to show the number of children in each group. Add numbers to the *y*-axis.

Aqua Swim School groups

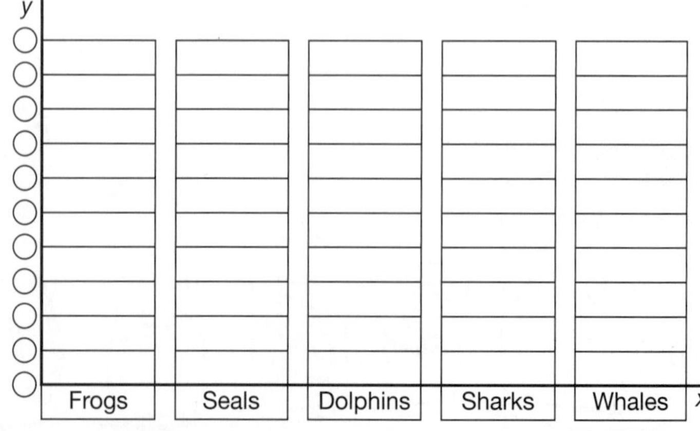

Number and Algebra

SET 3 Ordering decimals

Colour the largest measurement in each group.

1	0.8 m	0.83 m	0.38 m
2	0.1 m	0.01 m	0.11 m
3	0.31 m	3.1 m	3.01 m
4	8.07 m	0.87 m	7.08 m
5	6.09 m	7.89 m	0.99 m

Continue each sequence.

6	3.1	3.2	3.3	
7	8.12	8.14	8.16	
8	0.7	0.75	0.8	
9	3.25	3.50	3.75	
10	7.02	7.09	7.16	
11	0.78	0.9	1.02	
12	1.17	1.37	1.57	

SET 4 Extension

1. How much are 5 books at $7.50 each?
2. Add 10 000 to 54 962.
3. $\frac{3}{5}$ of 70
4. ($\frac{3}{4}$ of 36) $\times$ 9
5. Which is larger: 112 or $10 \times 10 \times 10$?
6. Round 10 961 to the nearest thousand.
7. Estimate an answer to 188 + 487.
8. Increase 9561 by four tens.
9. Average 12, 14, 18, 16
10. 26.8 + 2.1
11. 0.75 of a kilometre = ☐ m
12. Write 745 in words.
13. How much is 4.5 kg at $6 per kg?
14. Which is larger: 3.5 or $3\frac{3}{10}$?
15. Write the numeral for $(4 \times 10\,000) + (6 \times 1000) + (3 \times 100) + (9 \times 10) + 8$.

Working Mathematically

16 Jane had $456 in her purse but spent $45 on dinner, $126 on a new dress, $168 on an MP3 player and lost $50.

How much did she have left? ________

Measurement The tonne

Tick the box that best describes the mass of each item.

	Less than 1 t	About 1 t	More than 1 t
elephant			
25 children			
television			
car			
bulldozer			
refrigerator			

Number and Algebra

SET 1 Basic

1 15 + 4 + 1

2 18 – 6

3 7×5

4 $35 \div 7$

5 What is the sum of 9 and 5?

6 \$3.71 = ☐ c

7 150 – 6

8 Triple 8.

9 14 + 12 + 2

10 40 – 7

11 8×9

12 9^2

13 $56 \div 8$

14 Product of 8 and 6

15

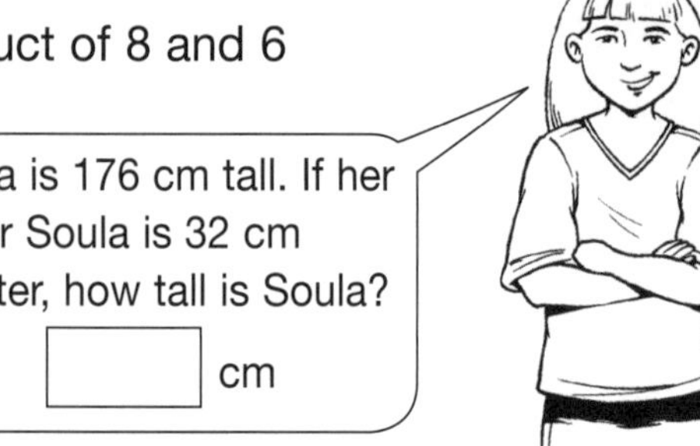

SET 2 Addition and estimation

Write 'reasonable' or 'unreasonable' to describe these estimates.

1	$31 + 73 \approx 100$	
2	$242 + 49 \approx 290$	
3	$315 + 480 \approx 800$	
4	$189 + 17 \approx 210$	
5	$573 + 425 \approx 1000$	
6	$163 + 115 \approx 275$	
7	$306 + 649 \approx 955$	
8	$1160 + 143 \approx 1200$	
9	$1701 + 82 \approx 1780$	
10	$818 + 138 \approx 960$	

11
$$\begin{array}{r} 23\,605 \\ +\ 6\,148 \\ \hline \end{array}$$

12
$$\begin{array}{r} 15\,125 \\ +\ 12\,306 \\ \hline \end{array}$$

13
$$\begin{array}{r} 38\,908 \\ +\ 10\,666 \\ \hline \end{array}$$

14
$$\begin{array}{r} 27\,080 \\ +\ 764 \\ \hline \end{array}$$

Space Plan view

The school needs to order new flooring for this classroom. Use the scale to calculate the square metres of carpet and lino needed.

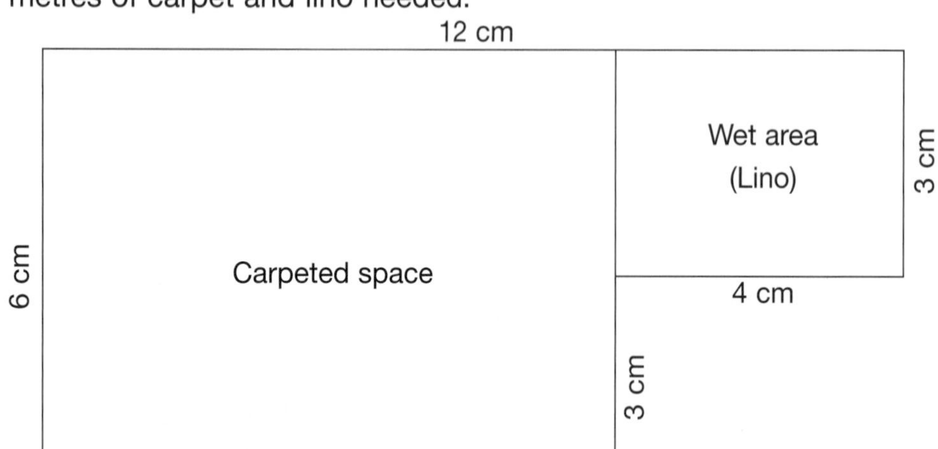

	Length	Width	Area
Carpet			
Lino			

Scale 1 cm = 1 m

Space

SET 3 Quadrilaterals

1 Colour the quadrilaterals.

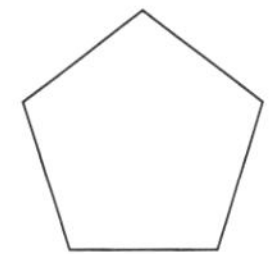

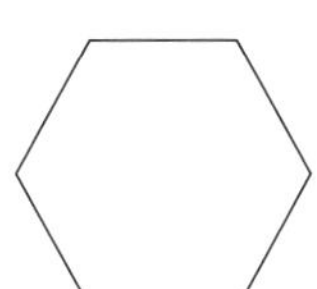

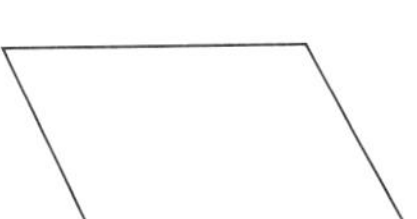
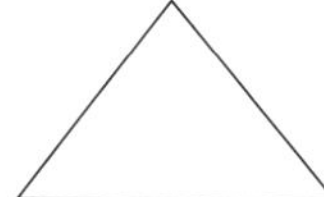

2 Colour the shape that matches the description.

I have 4 sides of equal length. I have 2 angles smaller than a right angle and 2 angles greater than a right angle.

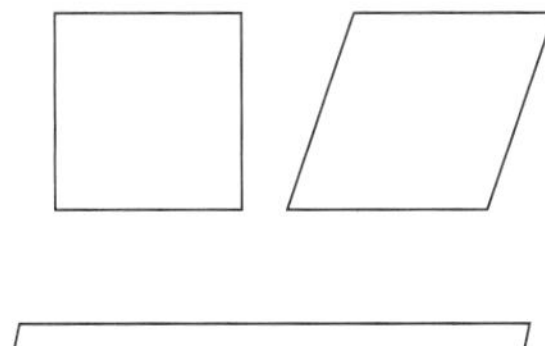

SET 4 Extension

1 2.6 + 3.3

2 Add 10 000 to 24 985.

3 (4 × 5) + (2 × 10)

4 How much is 4.5 kg at 80c per kg?

5 Which is larger: (4×10^2) or 4000?

6 Round 10 358 to the nearest ten thousand.

7 If 3 cost 66c, how much would 7 cost?

8 Decrease 7856 by four thousand.

9 Write a quarter to eleven in digital time.

10 Write ten thousand and six in numerals.

11 80% of a metre = ☐ cm

12 Which is larger: 7.23 or 3.99?

13 Round 6222 to the nearest hundred.

14 ($20 – 1) × 7

15 Write the numeral for (5 × 10 000) + (7 × 1000) + (4 × 100) + (0 × 10) + 9.

16 Write twenty-five past seven in the evening in 24-hour time.

17

Hanna	Sophie	Lisa	Grace	Zoe
149 cm	147 cm	154 cm	154 cm	146 cm

Which person's height is closest to the average for this group of girls?

Working Mathematically

Number and Algebra Number patterns with fractions

Use the number line to complete each pattern.

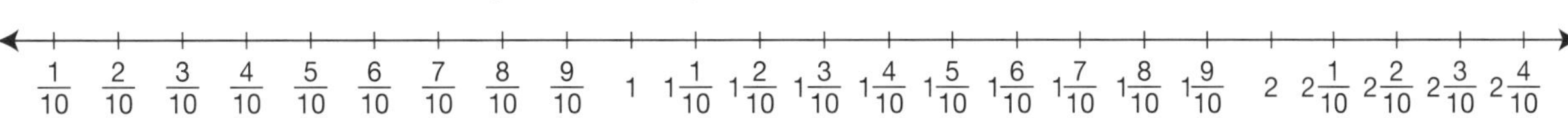

1 Add $\frac{3}{10}$

0	$\frac{3}{10}$	$\frac{6}{10}$					

2 Subtract $\frac{2}{10}$

2	$1\frac{8}{10}$	$1\frac{6}{10}$					

Number and Algebra

SET 1 Basic

1 17 + 8

2 27 ÷ 3

3 17 + 24

4 8 × 10

5 26 – 15

6 8 × 4

7 13 ÷ 6

8 What is the value of 6 in 36 705?

9 18 ☐ 18 = 36

10 46 ☐ 12 = 34

11 What is the sum of 16 and 22?

12 What is the product of 6 and 8?

13 Write twenty-seven in numerals.

14 Share $4 among 4 people.

15

Carlos has 27 football cards but swapped 7 of them for 2. How many does he have now?

☐ cards

SET 2 Multiplication problems

	One-way	Return
Sydney–Melbourne	$247	$494
Sydney–Broken Hill	$347	$694
Sydney–Brisbane	$265	$530

Calculate the cost of the trips.

1 2 return tickets to Melbourne

2 6 one-way tickets to Brisbane

3 4 return tickets to Broken Hill

4 8 one-way tickets to Broken Hill

5 Would you save money by buying a return ticket if you were coming back?

6 Mr Cook bought 4 return tickets that cost him $2120. What was his destination?

7 Ms Hill makes 3 return trips to Brisbane and 5 return trips to Broken Hill each year. How much does she spend?

Space Parallelograms

Draw and name the parallelogram that matches each description.

1	• four right angles • four sides of equal length • four axes of symmetry	
2	• four right angles • opposite sides are of equal length • two axes of symmetry	

Number and Algebra

SET 3 Decimals to thousandths

Use the symbols >, < or = to make each number sentence true.

1 9.351 ☐ 9.513
2 6.264 ☐ 6.624
3 5.781 ☐ 5.788
4 3.254 ☐ 3.255
5 8.972 ☐ 8.9
6 2.199 ☐ 2.19
7 9.254 ☐ 9.452
8 6.36 ☐ 6.360
9 5.202 ☐ 5.2
10 4.434 ☐ 4.044

Write the decimal before and after these numbers.

11 ________ (9.515) ________
12 ________ (4.398) ________
13 ________ (6.799) ________
14 ________ (8.400) ________
15 ________ (3.601) ________

SET 4 Extension

1 $9\frac{1}{2}$ m = ☐ cm
2 Write $4\frac{7}{100}$ as a decimal.
3 How many grams in 4.5 kg?
4 7.84 – 3.32
5 Perimeter of a square with sides of 14 cm
6 3.5 kg at $4 per kilogram
7 $2.75 × 6
8 3.15 + 2.07 + 3.1
9 What are the side lengths of a square if its area is 81 m^2?
10 Does a rectangular prism have an apex?

Working Mathematically

11 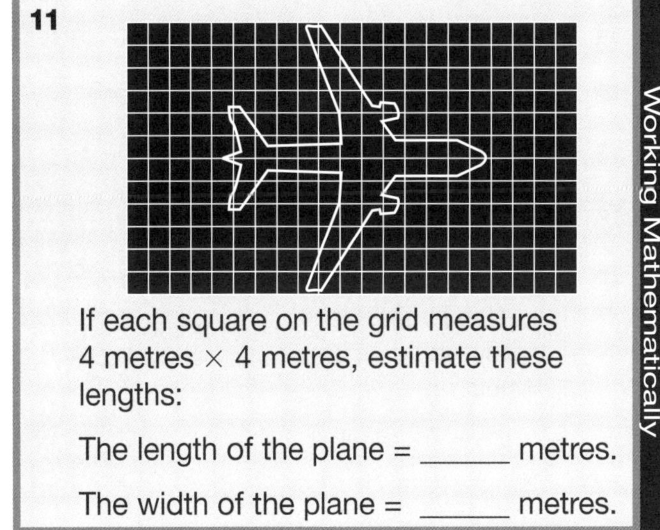

If each square on the grid measures 4 metres × 4 metres, estimate these lengths:

The length of the plane = ______ metres.

The width of the plane = ______ metres.

Measurement Timetables

Vegieville morning bus timetable

Carrot	Spud	Vegie	Pumpkin	Bean	Onion
5:55	5:58	6:02	6:10	6:20	
6:25	6:28	6:32	6:40	6:50	6:55
6:55	6:58	7:02	7:10	7:20	7:25
7:25	7:28	7:32	7:40	7:50	7:55
7:55	7:58	8:02	8:10	8:20	8:25
8:25	8:28	8:32	8:40	8:50	8:55
8:52	8:56	9:00	9:10	9:20	9:28
9:12	9:16	9:20	9:30	9:40	9:48
9:32	9:36	9:40	9:50	10:00	10:08
9:52	9:56	10:00	10:10	10:20	10:28
10:12	10:16	10:20	10:30	10:40	10:48
10:32	10:36	10:40	10:50	11:00	11:08
10:52	10:56	11:00	11:10	11:20	11:28
11:12	11:16	11:20	11:30	11:40	11:48
11:32	11:36	11:40	11:50	12:00	12:08
11:52	11:56	12:00	12:10	12:20	12:28

1 If you caught the 6:55 am from Carrot, when would you arrive at Onion? ______

2 If you caught the 7:02 am from Vegie, when would you arrive at Bean? ______

3 What bus would I need to catch from Spud to be at Onion by 10:48 am? ______

4 What bus would I need to catch from Vegie to be at Bean by 10:00 am? ______

5 How long does the 6:25 bus from Carrot take to reach Onion? ______

Number and Algebra

SET 1 Basic

1 7 ☐ 6 = 42

2 16 ☐ 4 = 20

3 What is the value of 1 in 10 374?

4 27 + 14

5 28 ÷ 7

6 17 + 7

7 9 × 6

8 8 × 9

9 43 – 8

10 24 ÷ 4

11 What is the difference between 30 and 17?

12 What is the 10th month of the year?

13 Share $54 among 6 people.

14 1248, 1240, 1232, ☐

15 Costa laid 24 bricks in 10 minutes. How many would he lay in 1 hour?

 bricks

SET 2 5-digit subtraction

Find the difference in price between these cars.

1

Van	
Car	

2

4WD	
Ute	

3

Station wagon	
Ute	

4

Ute	
Car	

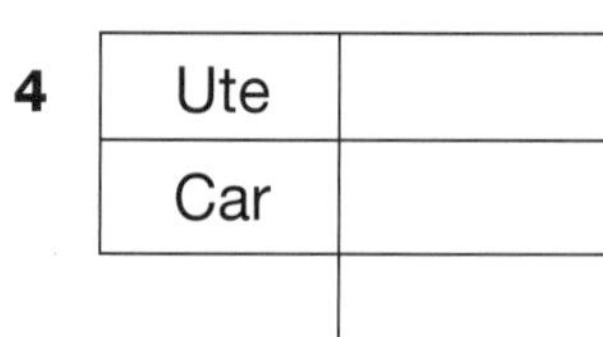

5

4WD	
Station wagon	

6

Station wagon	
Car	

Statistics and Probability Interpreting graphs

Miss Helga's class made a graph showing the different eye colours of the students.

1 Which eye colour was the most common? ___________

2 If there were 30 students in Miss Helga's class, how many had blue eyes? ___________

3 Was green the least common eye colour? ___________

4 Which was the most common eye colour out of brown and hazel? ___________

Miss Helga's class eye colours

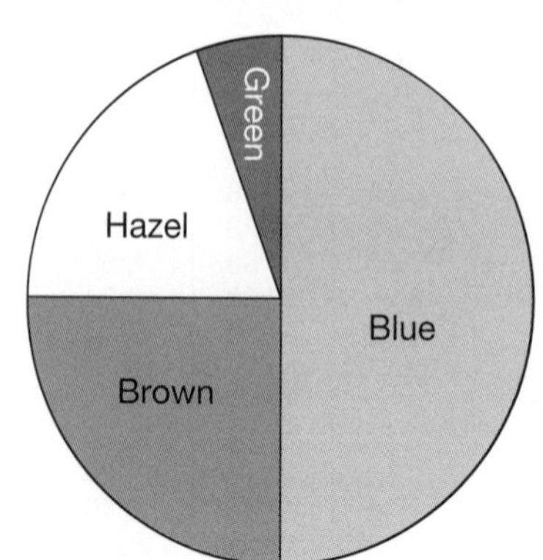

Number and Algebra

SET 3 Adding and subtracting fractions

1 $\frac{7}{10} + \frac{2}{10} =$

2 $\frac{8}{10} - \frac{5}{10} =$

3 $\frac{9}{10} - \frac{4}{10} =$

4 $\frac{7}{8} - \frac{3}{8} =$

5 $\frac{4}{8} + \frac{1}{8} =$

6 $\frac{4}{5} - \frac{1}{5} =$

7 $\frac{3}{5} + \frac{1}{5} =$

8 $1 - \frac{3}{5} =$

9 $\frac{2}{6} + \frac{3}{6} =$

10 $\frac{7}{12} - \frac{5}{12} =$

11 $\frac{1}{6} + \frac{4}{6} =$

12 $\frac{3}{12} + \frac{8}{12} =$

13 $\frac{2}{6} + \frac{2}{6} =$

14 $\frac{3}{8} + \frac{4}{8} =$

Mamma's slab pizza

15 If Tom ate $\frac{3}{18}$, Jerome $\frac{5}{18}$ and Sakiko $\frac{4}{18}$, how much would be left?

16 How could an 18-piece pizza be shared between two people so that one person receives twice as much as the other?

Working Mathematically

SET 4 Extension

1 Do the hands on a clock at 5 pm form an obtuse angle?

2 Write $14\frac{9}{100}$ as a decimal.

3 0.8 m = ☐ cm

4 Round 3.79 to the nearest tenth.

5 What is the radius of a circle if 10 cm is the diameter?

6 How much is 8 kg of meat at $6.75 per kilogram?

7 Hundredths in 1.76

8 Perimeter of an octagon with 8 cm sides

9 Which is largest: 33%, $\frac{30}{100}$, or 0.34?

10 8.6 m = ☐ cm

11 What fraction of a century is 25 years?

12 $22 + 3^2 + 4^2$

13 $\frac{2}{3}$ of $36

14 How much is $3\frac{1}{2}$ m of timber at $7.40 per metre?

15 Average of 92, 37, 101 and 10

16 If 2 bananas cost $0.90, how much would 9 bananas cost?

Space Three-dimensional views

Draw the top, front and side views of each shape.

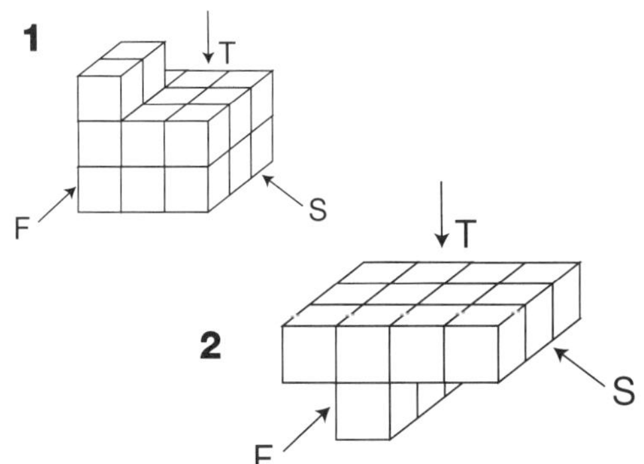

Top	Front	Side

UNIT 23

Number and Algebra

SET 1 Basic

1 10×5
2 $40 + 8$
3 $21 - 3$
4 $42 \div 6$
5 6×9
6 $40 + 7$
7 $35 - 6$
8 $18 \div 9$
9 What is the product of 4 and 0?
10 3×7
11 $25 + 4$
12 $19 - 8$
13 $36 \div 4$
14 What is the sum of 8, 6 and 9?
15

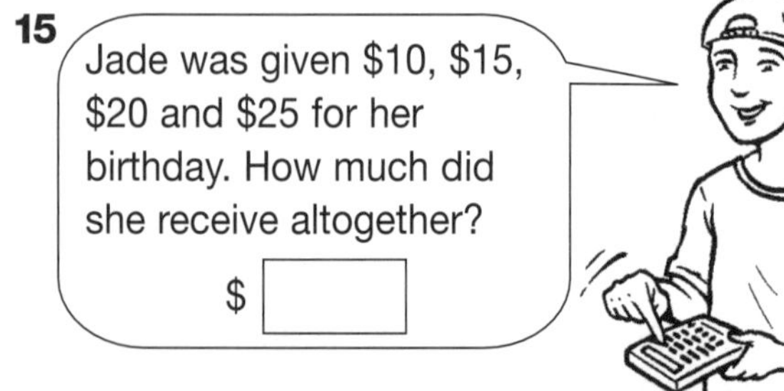

SET 2 4-digit multiplication

Round the 4-digit numbers to the nearest thousand to estimate these products.

1 $4895 \times 5 =$
2 $7106 \times 9 =$
3 $5987 \times 6 =$
4 $8214 \times 8 =$
5 $6037 \times 3 =$
6 $9320 \times 2 =$
7 $6988 \times 7 =$
8 $5895 \times 4 =$
9 $8171 \times 5 =$
10 $7202 \times 8 =$

Solve these multiplications.

11 4567×6

12 3894×8

13 How much money was collected if 5 people each paid $7654 for a trip?

Space Coordinates

Join the coordinate sets to form a shape.

1 Join (B,1) to (B,7).
2 Join (B,7) to (F,7).
3 Join (F,7) to (F,6).
4 Join (F,6) to (D,6).
5 Join (D,6) to (D,5).
6 Join (D,5) to (F,5).
7 Join (F,5) to (F,4).
8 Join (F,4) to (D,4).
9 Join (D,4) to (D,1).
10 Join (D,1) to (B,1).

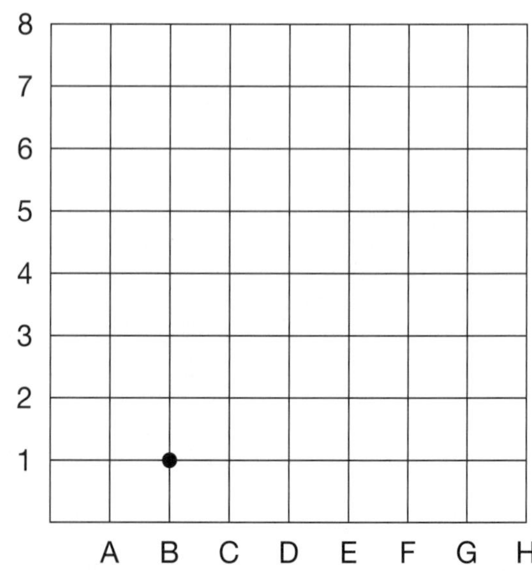

Number and Algebra

SET 3 Subtracting fractions from wholes

Subtract the fractions from the wholes.

1 $1 - \frac{1}{2} =$

2 $1 - \frac{1}{3} =$

3 $1 - \frac{1}{4} =$

4 $1 - \frac{1}{5} =$

5 $1 - \frac{1}{6} =$

6 $1 - \frac{1}{8} =$

7 $1 - \frac{1}{12} =$

8 $2 - \frac{1}{2} =$

9 $2 - \frac{1}{3} =$

10 $2 - \frac{1}{4} =$

11 $3 - \frac{1}{2} =$

12 $3 - \frac{1}{3} =$

13 $3 - \frac{1}{4} =$

14 $4 - \frac{1}{2} =$

Add or subtract the fractions.

15 $\frac{5}{8} + \frac{2}{8} =$

16 $\frac{3}{8} + \frac{4}{8} =$

17 $\frac{7}{12} + \frac{11}{12} =$

18 $\frac{11}{12} - \frac{7}{12} =$

19 $\frac{7}{8} - \frac{5}{8} =$

20 $\frac{4}{5} - \frac{3}{5} =$

21 Jim had a block of chocolate but gave away $\frac{3}{5}$ of it. How much did he have left? ☐

SET 4 Extension

1 $\frac{4}{5} + \frac{3}{5}$

2 What is the value of 7 in 37 829?

3 How many lines of symmetry has a rectangle?

4 4550 + 608

5 $(3 \times 10^2) + (4 \times 10) + 16$ ones

6 How much are 8 books at $2.80 each?

7 $(\frac{3}{5} \times 50) + (\frac{3}{8} \times 40)$

8 If 5 cost 95c, how much would 40 cost?

9 How much are 4 combs at $4.50 each?

10 10:25 pm + 35 minutes

11 Write two hundred and nineteen in numerals.

12 19 000 – 200

13 Write one hundred and eighty-four in Hindu–Arabic numerals.

14 What is the perimeter of a rectangle 18 cm long and 6 cm wide?

15 Write twelve thousand, six hundred and twenty-nine in Hindu–Arabic numerals.

16 Bill's mass is 48 kg. If Sam's mass is the same as Bill's plus $\frac{1}{3}$, what is Sam's mass?

Statistics and Probability Tree diagrams

Maya wanted a pizza for dinner and her choices were thick or thin, pepperoni or chicken, olives or mushrooms. Draw a tree diagram to show all possible combinations of pizza she could have.

UNIT
24

Number and Algebra

SET 1 Basic

1 15 ÷ 3

2 18 + 18

3 32 + 14

4 7 × 6

5 43 – 6

6 4 × 8

7 8 × 4

8 36 ☐ 8 = 44

9 8 ☐ 7 = 56

10 What is the value of 6 in 37 364?

11 What is the sum of 17 and 24?

12 What is the product of 6 and 9?

13 Write the factors of 100.

14 13 750, 13 755, 13 760, ☐

15

Thelma and Louise saved $27 and $46. How much more do they need to save to buy an $85.50 present?

$ ☐

SET 2 Division by 10

Solve these divisions.

1 10⟌360

2 10⟌420

3 10⟌650

4 10⟌700

5 10⟌434

6 10⟌480

7 10⟌358

8 10⟌900

9 10⟌862

10 10⟌940

11 10⟌880

12 10⟌575

13 10⟌390

14 10⟌595

15 10⟌936

16 Phillipa shared 180 football cards among herself and 9 others. How many did each person get?

17 Sophie saved $860 over ten months. What was her average saving per month?

Space Measuring angles

Calculate and record the size of each angle on the grid.
The total angle sum of a triangle is 180°.

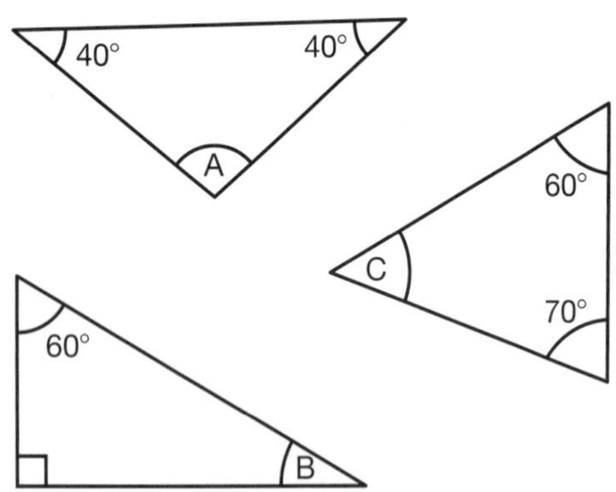

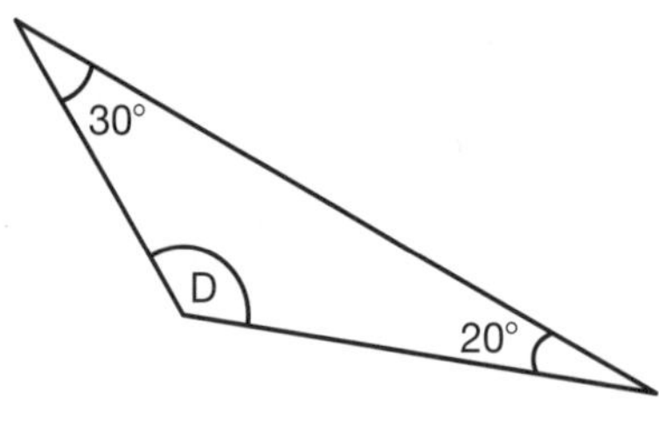

Angle	Degrees
A	
B	
C	
D	

Number and Algebra

SET 3 Fractions of a quantity

1 Find $\frac{1}{4}$ of 20.
2 Find $\frac{1}{5}$ of 20.
3 Find $\frac{1}{2}$ of 80.
4 Find $\frac{1}{3}$ of 27.
5 Find $\frac{1}{8}$ of 40.
6 Find $\frac{1}{10}$ of 100.
7 Find $\frac{3}{4}$ of 24 sheep.
8 Find $\frac{3}{5}$ of 50 goats.
9 Find $\frac{7}{10}$ of 30 cakes.
10 Find $\frac{4}{5}$ of 45 rabbits.

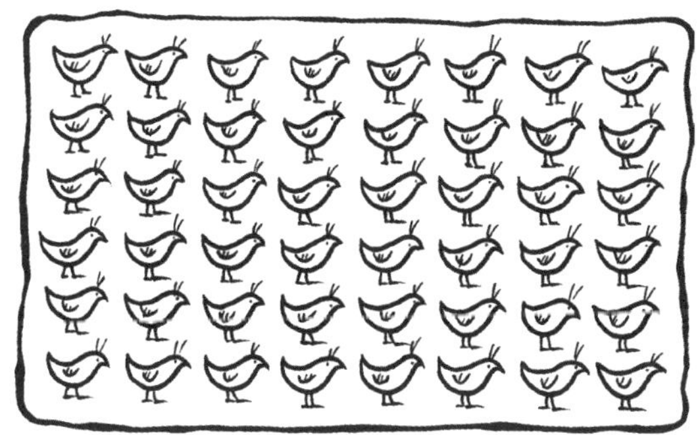

Find the fractions of the 48 birds.

11 $\frac{1}{4}$ =
12 $\frac{1}{8}$ =
13 $\frac{1}{2}$ =
14 $\frac{1}{3}$ =
15 $\frac{1}{6}$ =
16 $\frac{3}{4}$ =
17 $\frac{5}{8}$ =
18 $\frac{2}{3}$ =

SET 4 Extension

1 How many centimetres in $9\frac{1}{4}$ metres?
2 $(15 + 8) + 5^2$
3 If 9 pencils cost 72c, how much would 17 cost?
4 How many degrees in a rectangle?
5 An obtuse angle is less than 90°. True or false?
6 3, 6, 12, 24, ☐
7 At 3, John is 75 cm tall. If he grows $5\frac{1}{2}$ cm a year, how tall will he be when he's 11?
8 I had \$73 but spent \$43.50. How much money have I now?
9 What is the perimeter of a hexagon with 15 cm sides?
10 7 × 836
11 13 750c = \$ ☐
12 Round 31 326 to the nearest 10 000.
13 How many minutes from 10:43 am to 2:15 pm?
14 Average 43, 27, 51, 13 and 16
15 What is $\frac{4}{5}$ of \$90?
16 Write 10:45 pm in 24-hour time.

Measurement Time zones

Western Standard Time is $1\frac{1}{2}$ hours behind Central Standard Time. Central Standard Time is $\frac{1}{2}$ hour behind Eastern Standard Time. Use this information to find the times in the following capital cities.

Perth	Darwin	Brisbane	Sydney	Hobart	Melbourne	Adelaide
9 am				11 am		
		1:30 pm				
	12:15 pm					

Darwin
Western Standard
Central Standard
Eastern Standard
Brisbane
Perth
Adelaide
Sydney
Melbourne
Hobart

Number and Algebra

SET 1 Basic

1 18 + 3

2 17 – 13

3 13 + 16

4 9 × 4

5 16 – 12

6 9 × 6

7 45 ÷ 5

8 What is the value of 3 in 43 241?

9 8 ☐ 4 = 32

10 26 ☐ 17 = 9

11 What is the sum of 48 and 22?

12 What is the difference between 48 and 28?

13 How many days in August and September?

14 2 m = ☐ cm

15

SET 2 Calculator problems

Use your calculator's memory function to solve these problems.

1 Rayshaun bought 5 computers for $1987 each. How much did he spend? $ 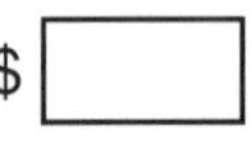

2 Jessica bought 5 CDs for $29 each and 3 books at $17 each. How much did she spend? $ 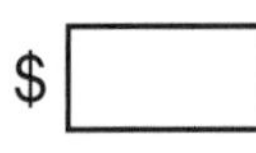

3 Thomas bought 2 model planes for $197 each and 6 packets of glue at $3.50. How much did he spend? $ 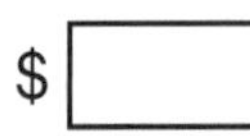

4 Kim saved $137 a month for 6 months and $209 a month for another 6 months. How much did Kim save? $ 

5 Solly wants a new TV set for $537. If she saves $47 a month for 7 months, how much more will she need? $ 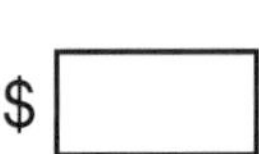

6 Peter bought a box of 48 apples for $16.80. How much was each apple? $

Number and Algebra Improper fractions

Use the shapes to add the fractions.

1

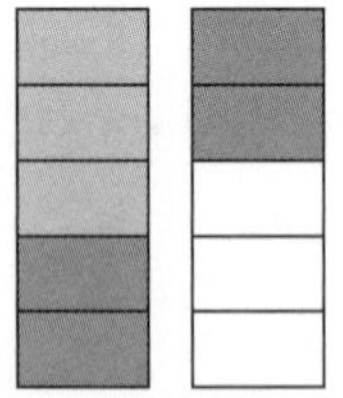

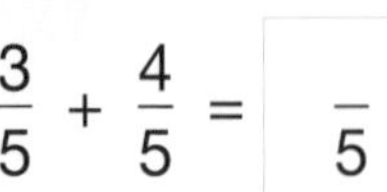

$\frac{3}{5} + \frac{4}{5} = \frac{}{5}$

2

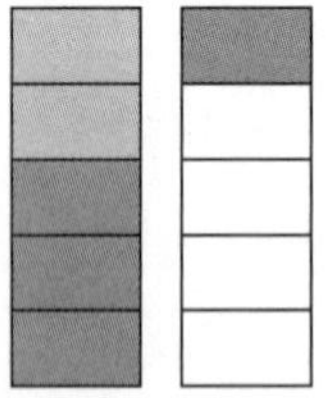

$\frac{2}{5} + \frac{4}{5} = \frac{}{5}$

Add these fractions. Record your answers as improper fractions.

3 $\frac{5}{10} + \frac{8}{10} =$

4 $\frac{6}{8} + \frac{3}{8} =$

5 $\frac{5}{8} + \frac{5}{8} =$

6 $\frac{3}{5} + \frac{4}{5} =$

7 $\frac{4}{10} + \frac{7}{10} =$

8 $\frac{8}{10} + \frac{3}{10} =$

9 $\frac{2}{4} + \frac{3}{4} =$

10 $\frac{3}{6} + \frac{4}{6} =$

11 $\frac{7}{10} + \frac{7}{10} =$

12 $\frac{9}{10} + \frac{8}{10} =$

13 $\frac{7}{10} + \frac{8}{10} =$

14 $\frac{8}{10} + \frac{8}{10} =$

15 $\frac{9}{10} + \frac{5}{10} =$

Number and Algebra

SET 3 Multiplication by tens

	×	10	100	1000
1	2	20		
2	4			
3	8			
4	11			
5	12			

6 $25 \times 20 =$ ______

7 $34 \times 30 =$ ______

8 $29 \times 40 =$ ______

9 $123 \times 50 =$ ______

10 $137 \times 60 =$ ______

11 $154 \times 70 =$ ______

12 Mr King bought 139 computers at an auction for $70 each. How much did they cost him?

SET 4 Extension

1 0.3 of 1 m = ☐ cm

2 What is one-third of 81?

3 An acute angle is less than 90°. True or false?

4 160, 80, 40, ☐

5 How many degrees between north and south?

6 Tim travelled for 6 hours at an average speed of 95 km/h. How far did he travel?

7 1350 + 4638

8 Round 3.69 to the nearest tenth.

9 How many eggs in $7\frac{1}{4}$ dozen?

10 The total mass of 6 men in a football scrum was 510 kg. What was the average mass?

Working Mathematically

11 How many people play baseball, soccer and hockey if 50 play basketball and 10 play tennis?

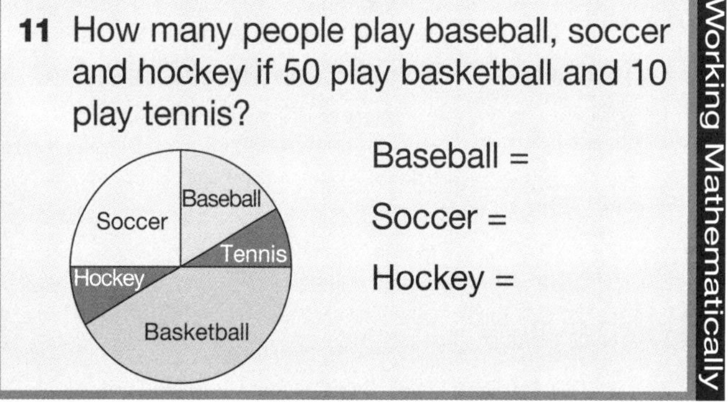

Baseball =

Soccer =

Hockey =

Measurement Gross mass/net mass

Complete the chart that shows the difference between the gross mass and the net mass of these items.

	1 Baked Beans	2 Honey	3 Potato Chips	4 Butter	5 Ice Cream
Gross	247 grams			375 grams	481 grams
Net	200 grams	500 grams	100 grams		
Difference		68 grams	72 grams	125 grams	81 grams

UNIT 26

Number and Algebra

SET 1 Basic

1 16 + 8

2 38 – 7

3 63 ÷ 7

4 What is the product of 8 and 7?

5 What is the sum of 9, 10 and 11?

6 4 + 4 + 6

7 27 – 8

8 20 ÷ 4

9 What is the average of 4, 6 and 5?

10 What is the difference between 21 and 3?

11 What is the cost of 4 books at $1.10 each?

12 400 cm = ☐ m

13 50 ÷ 10

14 60 – 40

15 Ricki puts 37 books on the library shelves every hour. If she works for 4 hours, how many books does she put on the shelves?

☐ books

SET 2 2-digit multiplication

1
  33
×16

2
  31
×53

3
  18
×75

4
  27
×33

5
  82
×21

6
  45
×42

7
  12
×27

8
  31
×22

9
  62
×99

10 There are 42 boxes of chocolates in the carton. Each box contains 24 chocolates. How many chocolates are there?

Space Nets

Circle the prism that belongs to the net.

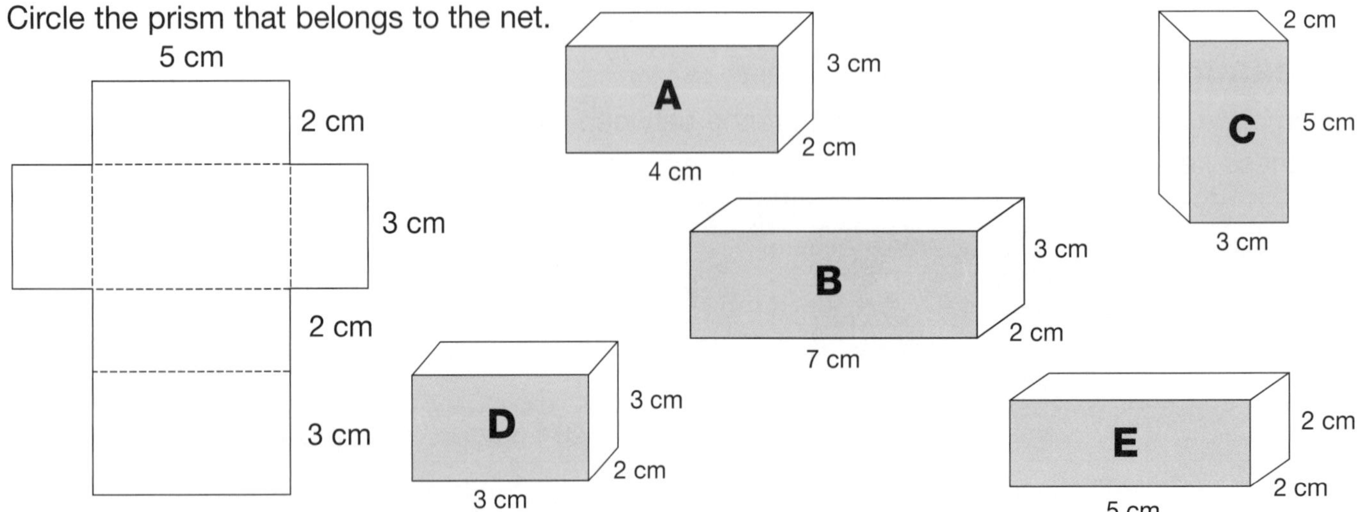

Number and Algebra

SET 3 Decimals

Write each fraction as a decimal.

1 $\frac{1}{10}$

2 $\frac{3}{10}$

3 $\frac{7}{10}$

4 $\frac{9}{10}$

5 $\frac{37}{100}$

6 $\frac{86}{100}$

7 $\frac{99}{100}$

8 $\frac{124}{1000}$

9 $\frac{189}{1000}$

10 $\frac{876}{1000}$

11 $\frac{777}{1000}$

12 $\frac{85}{1000}$

13 $\frac{67}{1000}$

14 $\frac{89}{1000}$

15 $\frac{7}{1000}$

Continue these decimal counting sequences.

16 45.321, 45.322, ______, ______, ______

17 27.987, 27.988, ______, ______, ______

18 79.547, 79.548, ______, ______, ______

19 68.888, 68.889, ______, ______, ______

20 35.107, 35.108, ______, ______, ______

SET 4 Extension

1 12 000 m = ☐ km

2 If a circle has a radius of 14 cm, what is its diameter?

3 How many thousands in 427 827?

4 What is the value of 6 in 5.06?

5 Round 16 295 to the nearest hundred.

6 If today is Tuesday 19 August, what is the date in 14 days' time?

7 ☐ + 16 = 4^2

8 A $250 bike is discounted by 10%. How much does it cost?

9 3.75 kg = ☐ g

10 75% of 200

11 30 L at $0.67 per litre

12 If 3 kg costs $54, how much would 8 kg cost?

Working Mathematically

13 US$1.00 = A$1.50

One American dollar is about equal to $1.50 Australian. What would be the value of US$10 in Australian dollars?

Statistics and Probability Line graphs

Mr and Mrs King collected growth data on their son Sam.

1 Create a line graph from the information below. (Age is given in years and height in centimetres.)

Age	1	2	3	4	5	6	7
Height	60	70	80	90	100	110	120

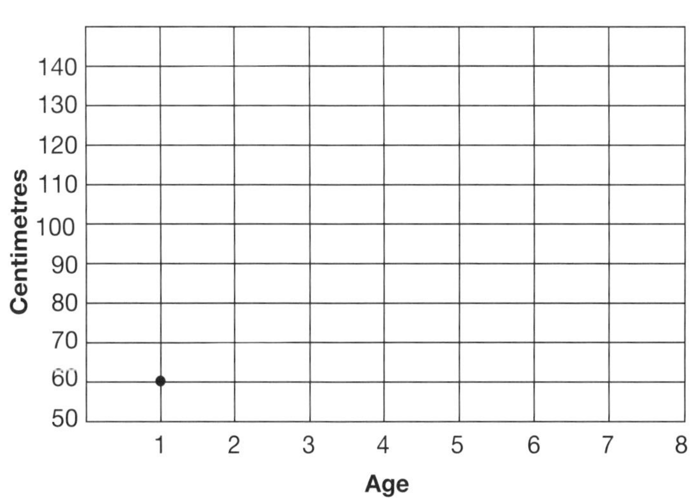

Answer the questions.

2 How tall was Sam at 4 years?

3 How tall was Sam at $6\frac{1}{2}$ years?

4 How old was Sam when he was 100 cm tall?

UNIT 27

Number and Algebra

SET 1 Basic

1 21 + 3 + 6

2 6×9

3 $42 \div 7$

4 What is the product of 7 and 4?

5 What is the sum of 9, 10 and 2?

6 $4.31 = ☐ c

7 What is the value of 4 in 34 329?

8 Share $16 among 4 people.

9 ☐ $\times 9 = 54$

10 What is the cost of 2 tickets at $1.20 each?

11 18×10

12 $(2 \times 4) + 5$

13 44 – 8

14 How many months are there in each season?

15

Mr Hill made $375 last week and $504 this week. How much did he make in the fortnight?

$ ☐

SET 2 Fractions of a quantity

1 $\frac{1}{4}$ of 36

2 $\frac{1}{5}$ of 35

3 $\frac{1}{8}$ of 32

4 $\frac{1}{10}$ of 60

5 $\frac{1}{4}$ of 28

6 $\frac{1}{3}$ of 27

7 $\frac{1}{6}$ of 42

8 $\frac{1}{2}$ of 100

9 $\frac{1}{8}$ of 40

10 $\frac{1}{10}$ of 70

Use the array below to find each fraction of the group of 24.

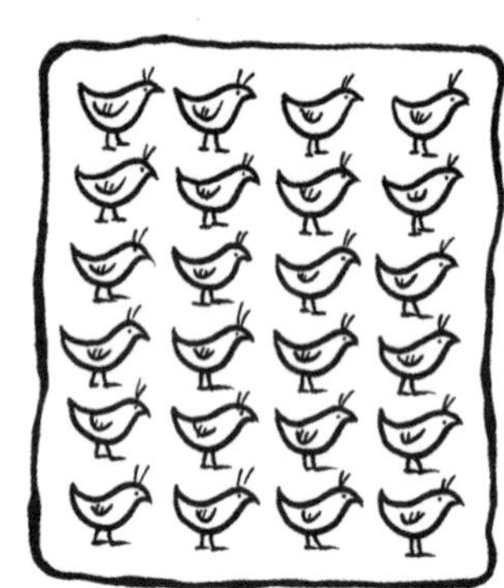

11 $\frac{1}{2}$ of 24

12 $\frac{1}{3}$ of 24

13 $\frac{1}{4}$ of 24

14 $\frac{1}{6}$ of 24

15 $\frac{1}{8}$ of 24

Measurement Cubic centimetres and millilitres

Complete the comparison chart to show equivalence. (1 cm^3 displaces 1 mL of water.)

Cubic centimetres	Millilitres
3 cm^3	mL
6 cm^3	mL
cm^3	12 mL
cm^3	24 mL
36 cm^3	mL
24 cm^3	mL

Cubic centimetres	Millilitres	Litres
cm^3	200 mL	
250 cm^3	mL	
cm^3	400 mL	
1000 cm^3	mL	1 L
cm^3	2000 mL	2 L
3000 cm^3	mL	3 L

Number and Algebra

SET 3 Division

1 $7\overline{)840}$

2 $6\overline{)618}$

3 $5\overline{)525}$

4 $8\overline{)832}$

5 $4\overline{)834}$

6 $6\overline{)630}$

7 $3\overline{)907}$

8 $9\overline{)936}$

9 $7\overline{)700}$

10 $8\overline{)700}$

11 $10\overline{)780}$

12 $10\overline{)960}$

13 Our house has an area of 238 m^2. If there are 7 rooms, what is the average size of each room?

14 Mohamed's group was given 840 blocks to share among their group of 8. How many did each person get?

15 $606 was shared among 6 winners. How much did each person receive?

SET 4 Extension

1 How much is 3.5 kg at $7 per kg?

2 Write 20 past 9 in digital form.

3 Round 17 653 to the nearest hundred.

4 60% of a metre = ☐ cm

5 Which is larger: (5×10^2) or 5000?

6 Write $15\frac{8}{10}$ as a decimal.

7 16 cm + 4 mm = ☐ mm

8 Hundredths in 4.64

9 How many 1.6 m lengths can I cut from an 8 m piece?

10 A cup holds 250 mL. How many would be needed to fill a $4\frac{1}{2}$ L jug?

11 How much is $4\frac{1}{2}$ m of ribbon at $3.20 a metre?

12 What is the average of 41, 49, 44 and 46?

Working Mathematically

13 Show all the possible totals when throwing two dice.

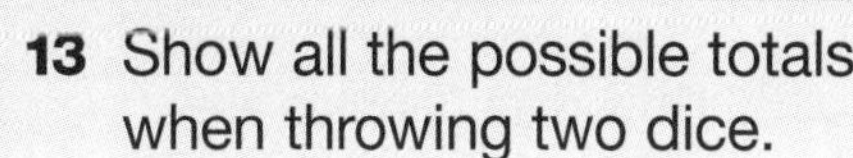

Dice	1	2	3	4	5	6
1	2	3	4	5	6	7
2	3	4	5	6		
3	4	5	6			
4	5	6				
5	6					
6						

Which total is the most common?

Measurement The hectare

Two soccer fields or a square measuring 100 m × 100 m equals 1 hectare. List items to match each heading in the grid.

More than 1 hectare	About 1 hectare	Less than 1 hectare

100 m

100 m

1 hectare
(10 000 m^2)

UNIT
28

Number and Algebra

SET 1 Basic

1 $5^2 + 3$

2 \$2.50 + 40c

3 3 m × 4

4 List the factors of 16.

5 Two places after 62nd

6 9 tens + 4 ones

7 How much is 6 m of material at \$20 per metre?

8 Divide 21 by 7.

9 95 × 0

10 \$4.10 + \$2.40

11 \$5.48 = ☐ c

12 8 × 0

13 What is half of 68?

14 72 = ☐ tens + ☐ ones

15

SET 2 4-digit multiplication

Round the 4-digit numbers to the nearest thousand to estimate these products.

1 6774 × 4 =

2 8914 × 9 =

3 9785 × 5 =

4 4576 × 8 =

5 3907 × 6 =

6 9379 × 3 =

7 8858 × 7 =

8 4653 × 4 =

9 7178 × 2 =

10 6999 × 8 =

Solve these multiplications.

11 7 5 5 8 × 8 = ____

12 9 4 3 2 × 6 = ____

13 How much money was collected if 5 people each paid \$6784 for a cruise?

Space Coordinates

1 Name the landmark at these coordinates.

a (3,6) ____________ b (7,2) ____________ c (17,7) ____________ d (20,3) ____________

2 Name any pair of coordinates for Deep Lake ____________ and Dark Forest ____________.

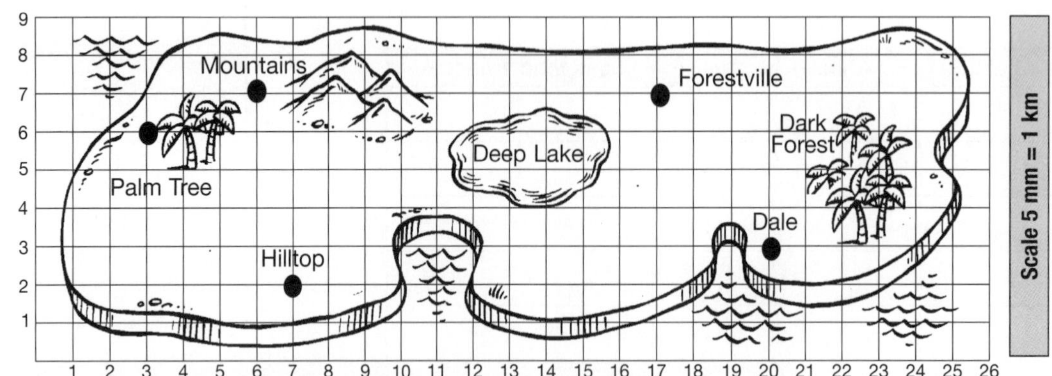

Number and Algebra

SET 3 6-digit addition

1
```
  424 655
+ 317 409
---------
```

2
```
  538 080
+ 424 533
---------
```

3
```
  643 720
+ 315 592
---------
```

4
```
  648 364
+ 239 783
---------
```

5
```
  855 555
+ 137 680
---------
```

6
```
  716 356
+ 263 562
---------
```

7
```
  316 341
  223 837
+ 217 601
---------
```

8
```
  521 007
  135 632
+ 212 555
---------
```

9 Sophia bought a car for $43 990, roof racks for $305, window tinting for $430 and a tow bar for $385. How much did she spend?

SET 4 Extension

1 What is the value of 6 in 3.61?

2 What is the perimeter of a 3.5 cm wide square?

3 Share $25 among 4 people.

4 6:51 pm + 18 minutes

5 Write sixty-four thousand, nine hundred and eighty-two in Hindu–Arabic numerals.

6 What direction is 135° clockwise from north?

7 How much is 250 g of mincemeat at $3.80 per kg?

8 If today is 27 September, what will be the date one fortnight from now?

9 4608 + 1300

10 $6.85 + $3.05

11 How many days in 3 years including a leap year?

12 A plane flew at 560 km/h for 3 hours. How far did it travel?

13 If Chan has a one-fifth share in a boat valued at $655, what is the value of his share?

14 One-half of a class of 32 are girls. If 9 girls don't like pizza, how many do?

Statistics and Probability Tabled data

Favourite subjects Y5/6

School	English	Maths	PE	Science
Mt Cook	4	8	3	10
Waverley	7	5	9	4
Tongarra	6	6	8	5
St Joseph's	9	6	7	3
Bond St	8	7	4	6
Broughton	7	5	8	5
St Paul's	10	6	2	7
Krambach	10	4	3	8

Twenty-five children from Years 5 and 6 were randomly selected to answer a favourite subject survey.

The results of the schools were recorded on a table.

Which school scored:

1 9 for English? ____________

2 8 for Maths? ____________

3 9 for PE? ____________

4 6 for Science? ____________

5 Overall, what was the most popular subject? ________

UNIT 29

Number and Algebra

SET 1 Basic

1 20 ÷ 4

2 34 + 16

3 17 – 9

4 8 × 9

5 43 – 16

6 37 + 37

7 8 × 4

8 16 ÷ 3

9 What is the value of 6 in 15 620?

10 16 ☐ 4 = 4

11 26 ☐ 24 = 50

12 How much is half of $46?

13 Triple 9.

14 120 minutes = ☐ hours

15 If today is Monday the 8th, what day will it be on the 18th of the same month? ☐

SET 2 Add and subtract decimals

Find the differences in prices between these items.

1

Bat	
Doll	

2

Guitar	
Blades	

3

Guitar	
Bat	

4

Skateboard	
Bat	

Find the total cost of these items.

5

Bear	
Doll	

6

Bat	
Bear	

Statistics and Probability Chance

1 Draw dots on the dice faces to show possible combinations of 7 when two dice are thrown.

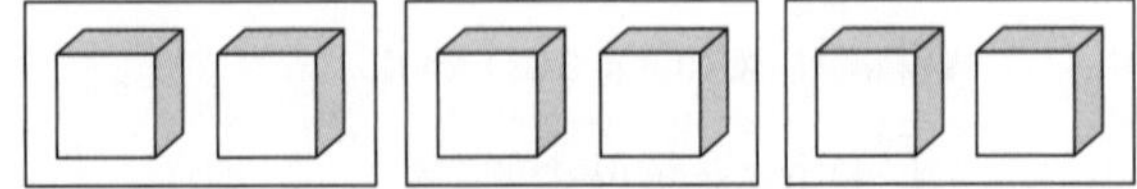

2 Draw dots on the dice faces to show possible combinations of 8 when two dice are thrown.

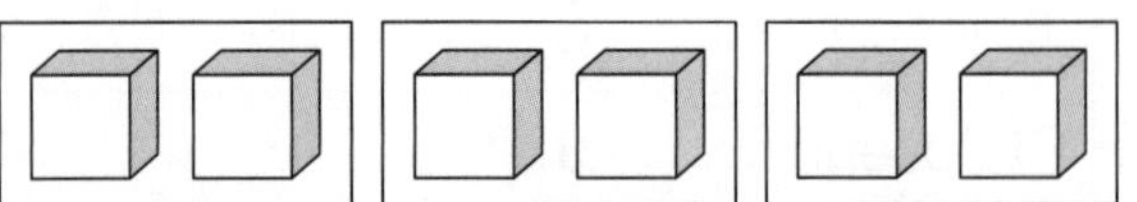

3 Draw dots on the dice faces to show possible combinations of 9 when two dice are thrown.

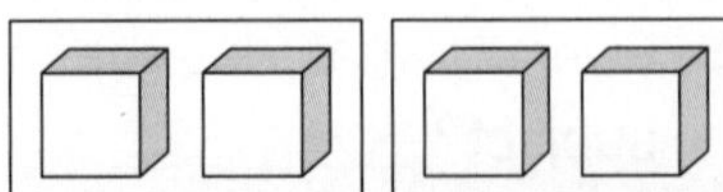

4 Draw dots on the dice faces to show possible combinations of 10 when two dice are thrown.

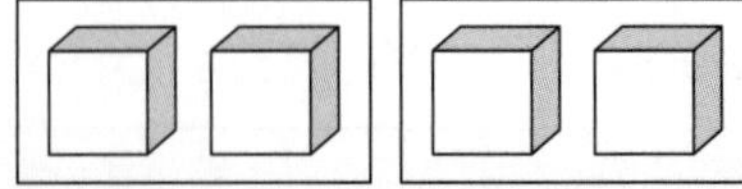

Number and Algebra

SET 3 Calculator problems

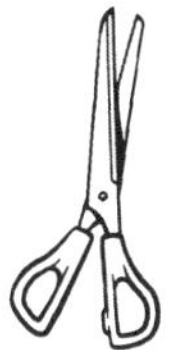
$3.55

$0.55

$2.85

$0.75

Use a calculator to solve the problems.

1	Tom bought 3 pairs of scissors and 10 pencils. How much did he spend?	
2	Gill bought 3 glue sticks, 4 rulers and 5 pencils. How much did she spend?	
3	Tony bought 6 rulers, 2 pencils and 7 pairs of scissors. How much did he spend?	
4	Tony spent $19.95 buying 7 lots of one of the items. Which item did he buy?	

5 a 
6.6 g

A $2 coin weighs 6.6 g. Calculate the mass of $10 worth of $2 coins.

b 
9 g

A $1 coin weighs 9 g. What would be the value of a stack weighing 54 g?

Working Mathematically

SET 4 Extension

1 Write $27\frac{43}{100}$ as a decimal.

2 If 8 cost $56, how much would 7 cost?

3 I had $100 but spent $43. How much have I now?

4 What is the perimeter of an equilateral triangle with sides of 13 cm?

5 If there are 180 legs and there are 40 dogs, how many people are there?

6 Round 13.71 to the nearest tenth.

7 Average 27, 43, 30 and 20

8 What fraction of 2 kg is 250 g?

9 How much is 8.5 kg of bacon at $5 per kilogram?

10 How many 125 mL containers would be needed to fill a container of 2.5 L?

11 If a 2.5 metre signpost has a shadow 5 metres long, how tall is the flagpole if its shadow is 8 metres long?

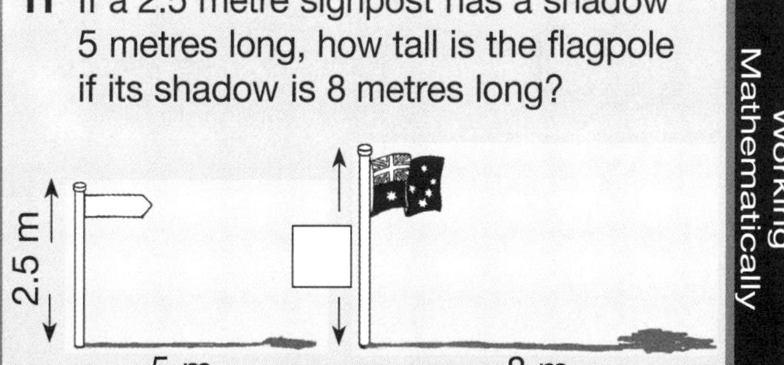

Working Mathematically

Space Nets

Colour the nets that will fold to make a cube.

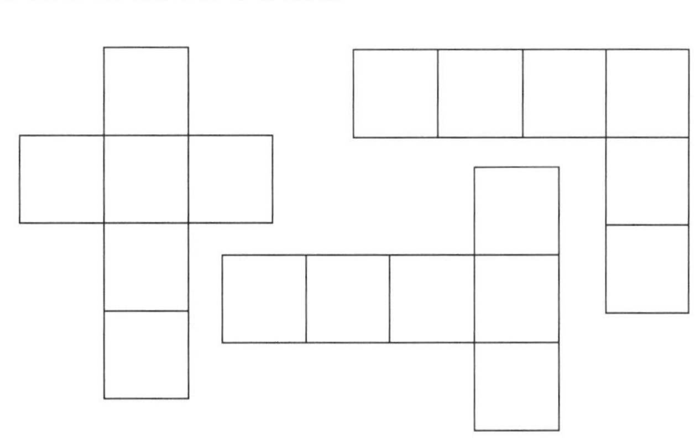

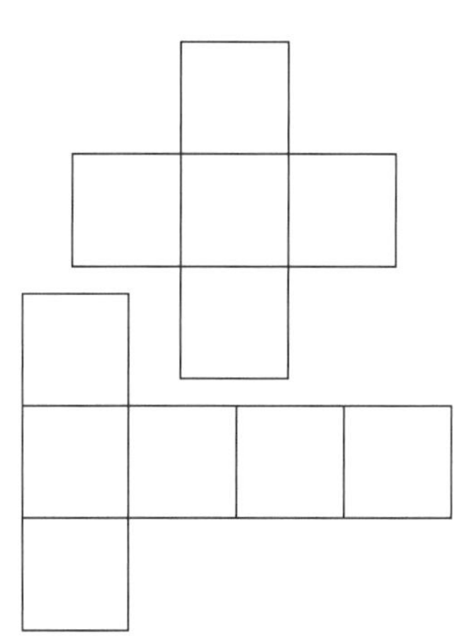

UNIT 30

Number and Algebra

SET 1 Basic

1 43 + 14

2 58 – 27

3 18 + 22

4 16 – 8

5 7 × 8

6 49 ÷ 7

7 64 ÷ 8

8 18 ☐ 3 = 6

9 17 ☐ 18 = 35

10 What is the product of 7 and 5?

11 Divide 36 by 6.

12 How many 5c coins in $1.15?

13 13 741, 13 751, 13 761, ☐

14 Write 90 in words.

15

Tim's mass is 54 kg and Kelly's is 36 kg. What is the difference in their weights?

☐ kg

SET 2 6-digit subtraction

1 $5127.56 − 4112.18

2 $7364.88 − 4158.93

3 854 800 − 237 283

4 816 395 − 7 648

5 983 754 − 355 555

6 850 907 − 349 013

7 Which two towns am I thinking of if the difference in population between them is 18 444?

Town	Population
Port Macquarie	33 709
Dubbo	30 102
Wyong	100 468
Wagga Wagga	42 848
Maitland	52 153
Orange	30 705
Lismore	27 246

Working Mathematically

Measurement Choosing units

Tick the box that displays the best measuring unit for each item.

	Item	mm	cm	m	km	g	kg	mL	L	ha
1	The length of a car									
2	The mass of a pencil									
3	The capacity of a drum									
4	The mass of a dog									
5	The capacity of a small cup									
6	The length of an ant									
7	The length of a ruler									
8	The area of a small farm									

Number and Algebra

SET 3 Using factors

Write all the factors for the following numbers. The first one is done for you.

1	10	1, 2, 5 and 10
2	12	
3	15	
4	16	
5	20	
6	24	

Separate the second multiple into factors to find the answer to the following.

E.g. 42×12 becomes $42 \times 4 \times 3 = 504$

7 23×12
8 31×15
9 41×12
10 52×15
11 43×14
12 39×12
13 40×20
14 32×16
15 44×16
16 33×20

SET 4 Extension

1 Area of a square with sides of 4 cm
2 How many 400 g packets in 2 kg?
3 $\frac{1}{4}$ of 2 km
4 1110, 1133, 1156, ☐
5 Value of 7 in 173.59
6 What fraction of $36 is $9?
7 Round 37 411 to the nearest hundred.
8 What is the perimeter of a decagon with sides of 17 cm?
9 Write seventy-seven in numerals.
10 $7\frac{1}{4}$ kg of sugar at $1.60 per kilogram.
11 Write $17\frac{17}{100}$ as a decimal.
12 3 books at $7.20 each.
13 1, 3, 9, 27, ☐
14 Months in $4\frac{1}{2}$ years
15 What angle is formed by the hands of a clock at 6 o'clock?
16 If 7 cost $42, how much would $7\frac{1}{2}$ cost?
17 $60 × $80
18 Round to the nearest $10 to answer $157.60 + $83.20.

Space Triangles and parallelograms

Draw a line to match the shapes to their names.

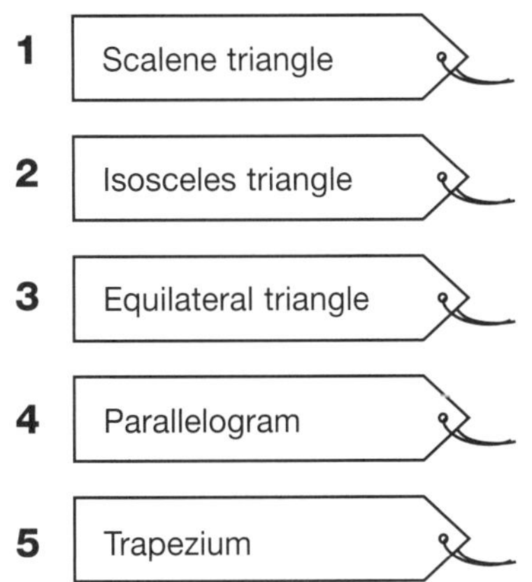

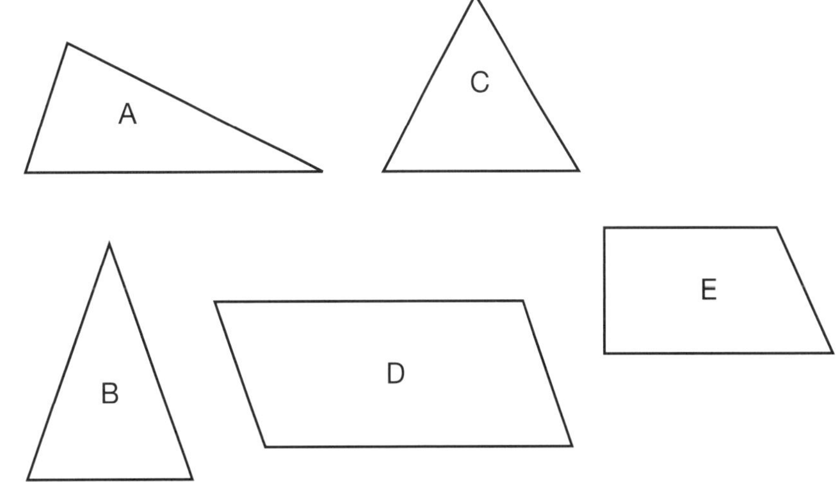

Number and Algebra

SET 1 Basic

1 25 ÷ 5

2 6 × 3

3 ☐ tens + ☐ ones = 63

4 300 + 40 + 5

5 12c + 20c

6 4 cm = ☐ mm

7 48 ÷ 8

8 $1.95 = ☐ c

9 $11 × 10

10 $21 – $19

11 400 + 60 + 6

12 4000 m = ☐ km

13 How far is one quarter of 20 km?

14 54 ÷ 9

15 

How many 5-kg bags of sugar can be filled from an 80-kg drum?
☐ bags

SET 2 Price per unit/best value

Complete the table to show the price per unit.

	Quantity and price	Unit price
1	3 kg for $21	$
2	5 kg for $45	$
3	4 kg for $32	$
4	6 kg for $72	$
5	8 kg for $56	$
6	10 kg for $35	$

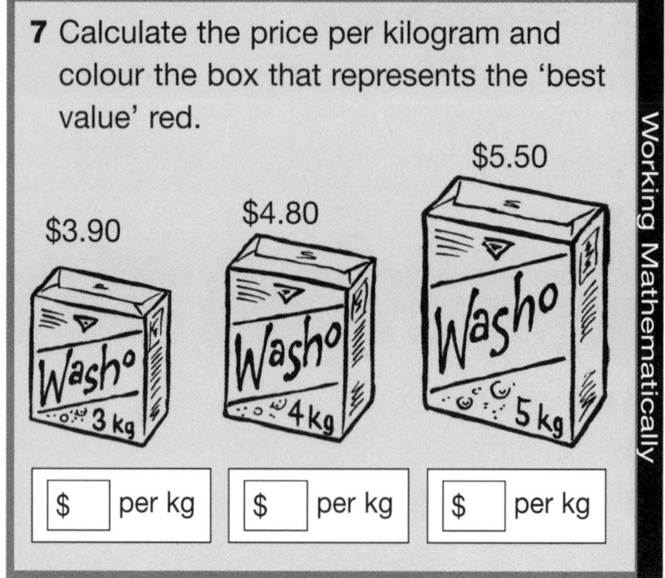

Measurement Area of triangles

Find the area of the rectangle, then halve it to find the area of the triangle.

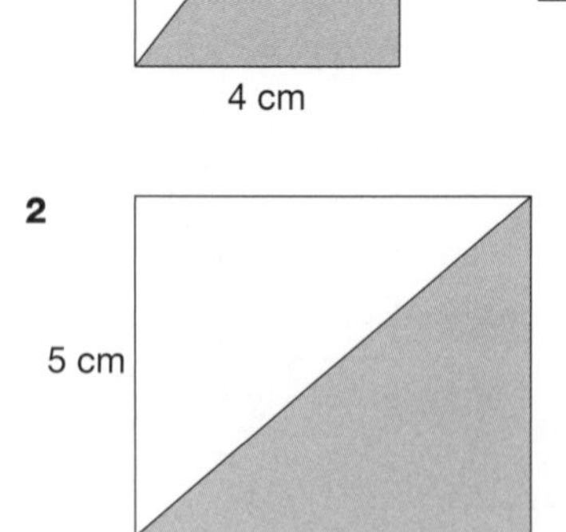

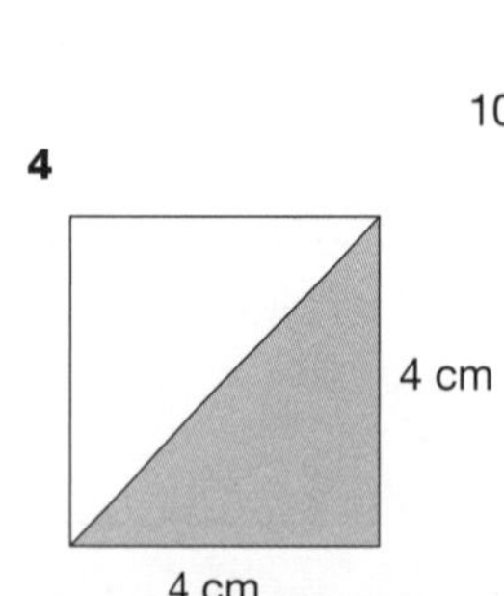

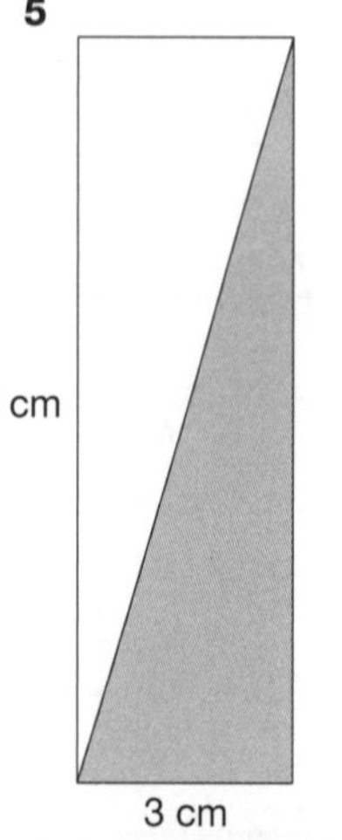

	Area of rectangle	Area of triangle
1		
2		
3		
4		
5		

Number and Algebra

SET 3 Inverse operations

Use an inverse operation to check each equation. The first one is done for you.

1	24 ÷ 3 = 8	✓	8 × 3 = 24
2	42 ÷ 7 = 7		
3	54 ÷ 7 = 8		
4	64 ÷ 8 = 8		
5	63 ÷ 9 = 7		
6	100 – 77 = 23		
7	200 – 137 = 63		
8	93 – 48 = 45		
9	200 – 37 = 53		

Working Mathematically

Check each subtraction with an addition.

10	11	12
837	4 623	2 796
– 123	– 135	– 199
714	4 488	2 593
714	4 488	2 593
+ 123	+ 135	+ 199
____	____	____

SET 4 Extension

1 How much are 7 pencils at 2 for 30c?

2 740 × 3

3 What is the value of 6 in 136.75?

4 14 500 + 300 + 20

5 1.25 kg at $2 per kilogram

6 $6.72 – $2.48

7 How much is 75 cm of tape at $2 per metre?

8 $(7^2 - 25) + (\frac{3}{5} \text{ of } 80)$

9 If 6 pencils cost $1.20, how many could I buy for $4?

10 $\frac{7}{10} + \frac{2}{5}$

11 If the perimeter of a rectangle is 84 cm and one side is 24 cm long, what is the length of the other side?

12 How many fifths in 7.2?

13 Jacqui got 60% of her spelling test correct. If there were 100 words, how many did she get right?

14 What is the product of (4×10^2) and 2?

15 What is the difference between 10^2 and 4^2?

16 How many prime numbers are there between 20 and 40?

Measurement Square kilometres

Tick true or false for each statement.

	Statement	True	False
1	The area of a room would be measured in hectares.		
2	The area of Sydney would be measured in square kilometres.		
3	The area of a soccer field would be about 1 hectare.		
4	1000 m × 1000 m = 1 square kilometre		
5	A noticeboard would be approximately 1 square metre.		
6	The area of Tasmania would be measured in square metres.		

UNIT 32

Number and Algebra

SET 1 Basic

1 700 ☐ 350 = 350
2 8 × 4
3 27 – 14
4 48 ÷ 6
5 17 + 18
6 15 ☐ 3 = 5
7 15 × 3
8 47 – 8
9 42 ÷ 7
10 36 301, 36 307, 36 313, ☐
11 What is the product of 7 and 3?
12 What is the sum of 47 and 26?
13 How much is half of 46?
14 Triple 6
15

Oranges cost $6 a dozen. How much do 36 oranges cost?

$ ☐

SET 2 Multiplication

1

×	10	100
3		
4		
7		
8		
5		

2

×	7	70
2		
4		
5		
3		
6		

3 257×7

4 $3\,254 \times 6$

5 $5\,792 \times 7$

6 27×35

7 325×7

8 258×8

9 Jim saved an average of $387 per month for 8 months. How much did he save?

Space Coordinates

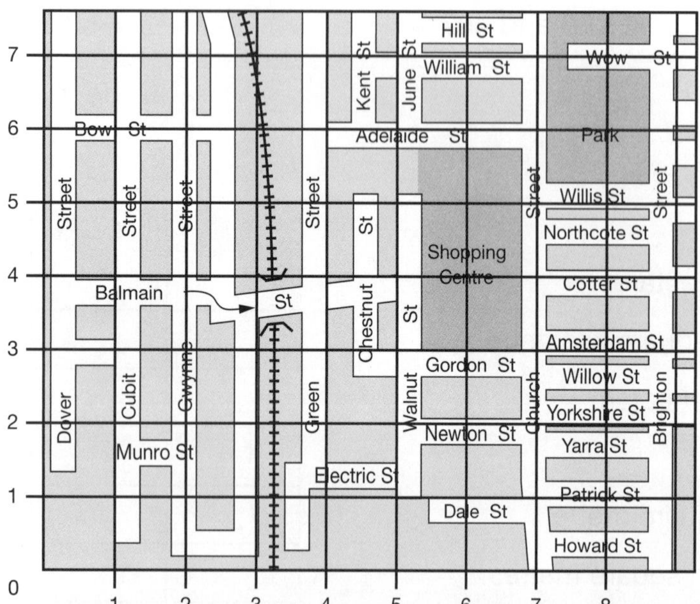

Find what is at these coordinates.

1 (6,2) ____________

2 (5,6) ____________

3 (8,4) ____________

4 (2,6) ____________

5 (8,7) ____________

6 (8,1) ____________

Number and Algebra

SET 3 Rounding numbers

Round these numbers.

	Number	Nearest 1000	Nearest 100	Nearest 10
1	1089	1000	1100	1090
2	2189			
3	3588			
4	4768			
5	5217			

6 Round Sydney's population of 4 627 345 to the nearest million ______________.

Working Mathematically

Round the numbers being divided to the nearest 10 or 100 to give quick estimations of the answers to these divisions. Then find the real answer.

	Number	Rounded to	Divided by	Estimate	Answer
7	397	400	4	100	$99\frac{1}{4}$
8	789		8		
9	791		4		
10	149		5		
11	254		5		

SET 4 Extension

1 How long is 4 decades minus 31 years?

2 What fraction of 1 minute is 45 seconds?

3 What fraction of 200 is 40?

4 Share 143 lollies between 6 people.

5 What is the perimeter of a rectangle with length 36 cm and width 23 cm?

6 $(\frac{3}{4} \times 32) \times (\frac{1}{5}$ of $40)$

7 Perimeter of a pentagon with 25 cm sides

8 How much is 4.25 kg of plaster at $16 per kilogram?

9 How much are 9 books at $4.05 each?

10 Which is the smallest: 10%, 0.1 or 0.09?

11 How many millilitres in 8.9 litres?

12 How many degrees in 2 triangles?

13 Write $27\frac{3}{100}$ as a decimal.

14 What is the area of a rectangle with length of 40 cm and width of 7 cm?

15 $8.75 = 8\frac{3}{4}$. True or false?

16 Add all prime numbers between 6 and 20.

Measurement Stopwatches

Study the grid before answering these questions.

1 Who won the race?

2 Who came last?

3 Who came third?

4 Who came second last?

5 Who came fifth?

6 By how many seconds did Sarah beat Alicia?

1500 m Race for 12-year-olds

Name	Time
Aaron	6 : 23 : 13
Chad	5 : 59 : 14
Sarah	6 : 05 : 67
Lorinne	5 : 59 : 35
Alicia	6 : 47 : 67
Christina	7 : 15 : 91

UNIT 33

Number and Algebra

SET 1 Basic

1 18 + 23

2 32 − 12

3 18 ☐ 2 = 36

4 18 ÷ 4

5 8 × 9

6 32 ☐ 36 = 68

7 45 ÷ 5

8 7 × 9

9 Write seventy-three in Hindu-Arabic numerals.

10 What is the difference between 43 and 15?

11 2 L = ☐ mL

12 How many tenths in $1\frac{1}{2}$?

13 How many months in spring and summer?

14 Write the set of factors for 12.

15

SET 2 Financial plans

	Fit Bodz	Muscles	Gymbo
Single visit	$5	$10	$5
5-visits	$20	$15	$20
10-visits	$30	$30	$40
20-visits	$60	$50	$80

Calculate the cost **per visit** to the different gyms.

1 Jon bought a 5-visit pass to Fit Bodz Gym. ______

2 Ali bought a 10-visit pass to Muscles Gym. ______

3 Ben bought a 20-visit pass to Gymbo Gym. ______

4 Alex bought a 10-visit pass and 20-visit pass to Fit Bodz. ______

5 Bec bought a 5-visit pass and a 10-visit pass to Muscles Gym. ______

Working Mathematically

Calculate the cheapest cost per visit for 30 sessions at:

Fit Bodz ______

Gymbo ______

Statistics and Probability Collecting data

Mr Peters' class collected data about age and long-jump ability.

Name	Age	Long jump
Josh	10	2.1 m
Con	10.5	2.15 m
Janice	11	1.9 m
Toula	11.6	2.5 m
Jeremy	11.8	2.4 m
Sally	12.1	2.1 m
Ophra	12.6	2.35 m

Study the table then answer the questions.

1 Did the oldest person jump the longest?

2 Did the youngest person jump the shortest?

3 List some things that may affect a person's long-jump ability.

______

______

______

Number and Algebra

SET 3 Percentages, fractions and decimals

Add the missing numbers to the chart.

	Fraction	Decimal	%
1	$\frac{10}{100}$		
2		0.2	
3			25%
4	$\frac{8}{10}$		
5	$\frac{1}{4}$		
6	$\frac{5}{10}$		
7			75%

Working Mathematically

Order from smallest to largest.

8	$\frac{25}{100}$	40%	0.04	
9	45%	$\frac{50}{100}$	0.54	
10	60%	0.06	$\frac{66}{100}$	
11	0.45	0.5	$\frac{40}{100}$	
12	5%	$\frac{5}{10}$	0.1	
13	$\frac{10}{100}$	0.7	11%	
14	$\frac{11}{100}$	10%	0.01	

SET 4 Extension

1 2.7 L = ☐ mL
2 Are 27 and 63 multiples of 9?
3 7.25 m = ☐ cm
4 $\frac{6}{8} + \frac{3}{4}$
5 $(307 \times 8) - 4$
6 Write $23\frac{7}{10}$ as a decimal.
7 How much is 5.25 kg of meat at $6 per kilogram?
8 $(\frac{3}{4} \times 72) \times (\frac{3}{8} \times 24)$
9 If 9 books cost $63, how much would 27 cost?
10 Round 43.99 to the nearest whole number.
11 List the factors of 48.
12 What is the area of a rectangle with sides 14 cm and 5 cm?
13 $(\frac{3}{5} \times 150) - (\frac{3}{4} \times 44)$
14 How many 600 mL bottles are needed to fill a 3 L container?
15 Average 174, 326 and 250
16 One-quarter of 60 trees are eucalypts. How many are not?

Space Enlarging area

Use the dot paper to enlarge each shape by doubling its dimensions. Starting points have been given.

1

2

UNIT 34

Number and Algebra

SET 1 Basic

1 54 ÷ 9

2 6 × 5

3 77 = ☐ tens + ☐ ones

4 400 + 60 + 5

5 48c + 6c

6 6 cm = ☐ mm

7 49 ÷ 7

8 $22.72 = ☐ c

9 $65 + $2.30

10 18 × 0

11 1600 + 40 + 8

12 One quarter of 28

13 72 ÷ 8

14 5c × 100

15 

There are 44 bricks in a square metre. How many in a 10 m^2 wall?

☐ bricks

SET 2 Percentages

1 10% of $200

2 10% of $400

3 10% of $800

4 10% of $1000

5 25% of $1000

6 25% of $120

7 50% of $600

8 50% of $800

9 1% of $1000

10 100% of $5000

Working Mathematically

BB3 hardware store is reducing the price of all their goods by 10%.

Complete the grid to show the discount amount and the new price.

	Original price	10% discount	New price
11	$500		
12	$700		
13	$2000		

Space Cross-sections

Draw the cross-section of each shape.

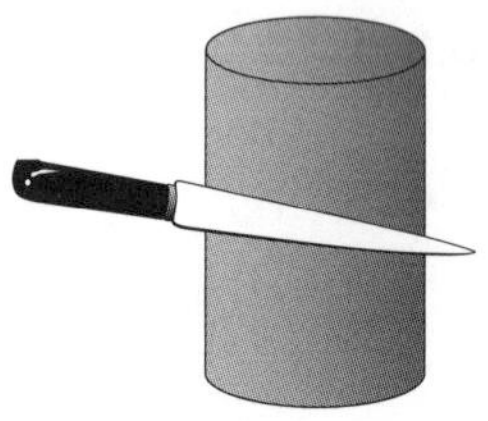

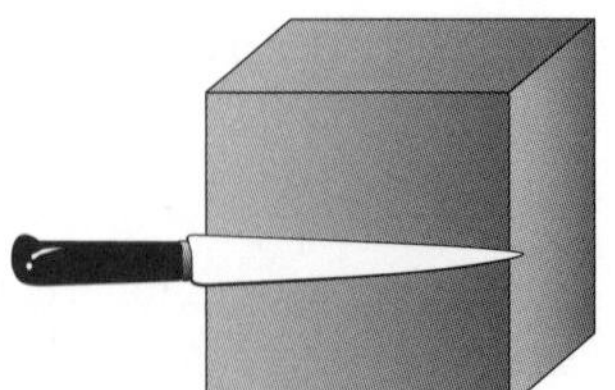

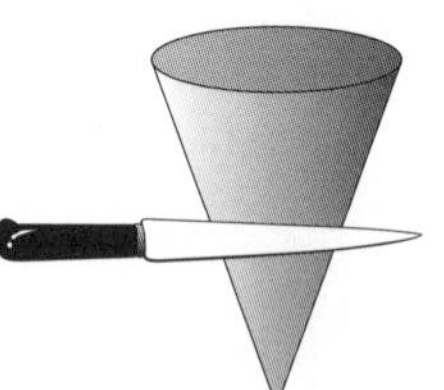

1	2	3	4

Number and Algebra

SET 3 Division with fractional remainders

Crack the code.

Remember: $\frac{2}{4}$ is the same as $\frac{1}{2}$.

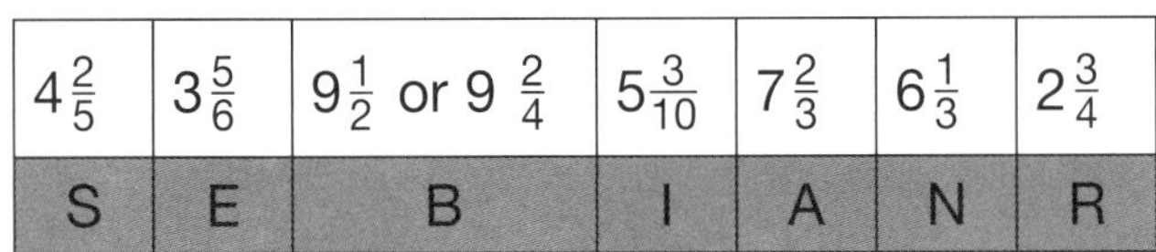

$4\frac{2}{5}$	$3\frac{5}{6}$	$9\frac{1}{2}$ or $9\frac{2}{4}$	$5\frac{3}{10}$	$7\frac{2}{3}$	$6\frac{1}{3}$	$2\frac{3}{4}$
S	E	B	I	A	N	R

1 $2\overline{)19}$

2 $4\overline{)11}$

3 $10\overline{)53}$

4 $5\overline{)22}$

5 $4\overline{)38}$

6 $3\overline{)23}$

7 $3\overline{)19}$

8 $6\overline{)23}$

1	2	3	4	5	6	7	8

9 If the 14 km Fun Run is divided into 4 equal sections, how many kilometres are in each section?

SET 4 Extension

1 What is the value of 7 in 78.65?

2 Find the perimeter of a triangle with all sides 8 cm.

3 Share $960 among 4 people.

4 Round 14.84 to the nearest tenth.

5 $180 - (\frac{7}{10}$ of $90)$

6 Minutes between 10:05 am and 11:49 am

7 25% of $900

8 How much is $8\frac{1}{2}$ m of timber at $1.90 a metre?

9 How much is 0.8 m of ribbon at $2 a metre?

10 What is the difference between 471 and 85?

11 How much are 7 trees at $9.75 each?

12 Round 14 870 to the nearest 1000.

13 $(\frac{1}{3}$ of $21) \times (\frac{7}{10}$ of $120)$

14 How many tricycles will 381 wheels go on?

15 What is the value of 6 in 79.68?

16 $50 \times 40 + 10^2$

Statistics and Probability Different graph displays

Woodstock Afterschool

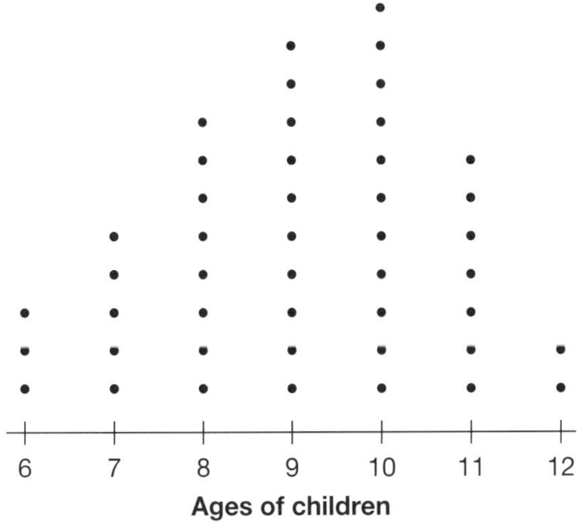

Study the dot plot graph to answer the questions.

1 Did more 11-year-olds attend than 7-year-olds?

2 Did more 9-year-olds attend than 10-year-olds?

3 What age group is the least represented?

4 How many people attend Woodstock Afterschool?

UNIT 35

Number and Algebra

SET 1 Basic

1 $35 \div 5$

2 $63 + 26$

3 $37 - 9$

4 $43 + 27$

5 $87 - 36$

6 9×4

7 8×7

8 $(3 + 6) \times 9$

9 What is the value of 7 in 37 306?

10 17 ☐ 3 = 51

11 87 + ☐ = 100

12 How many minutes in $3\frac{1}{2}$ hours?

13 Divide 50 by 4.

14 97 006 = ☐ + ☐ + ☐

15

How many 100 mL bottles are needed to fill a 2-L jug? ☐

SET 2 Equivalent number sentences

Circle the number that balances each equation.

	Equation	Possible solutions		
1	$15 \times 3 = \square + 8$	35	36	37
2	$10 + (\square \times 9) = 55$	5	10	15
3	$\square \div 9 \times 5 = 20$	45	36	54
4	$42 \div (3 + \square) = 6$	3	4	6
5	$81 - (\square \times 9) = 9$	8	9	72
6	$120 \div 10 \times \square = 12$	10	12	1
7	$54 \div 6 = \square \div 5$	44	45	89
8	$51 \div \square = 4^2 + 1$	17	3	13

Use = or ≠ to complete these equations.

9 $80 × 3 ☐ $60 × 4

10 1.5 kg × 3 ☐ 0.5 kg × 6

11 $0.90 × 6 ☐ $1.80 × 3

12 28 kg × 2 × 6 ☐ 9 kg × 5

13 2.5 m + 2.4 ☐ 0.7 m × 7

14 1500 mL × 500 ☐ 500 mL × 5

15 $21.90 ÷ 3 ☐ 73c × 10

Measurement Measurement units

Write a measurement unit you would use to measure the items.

1 The length of your finger ________

2 The length of a flea ________

3 The height of a door ________

4 The length of a highway ________

5 The capacity of a small cup ________

6 The volume of a sultana box ________

7 The mass of a pencil ________

8 The mass of a dog ________

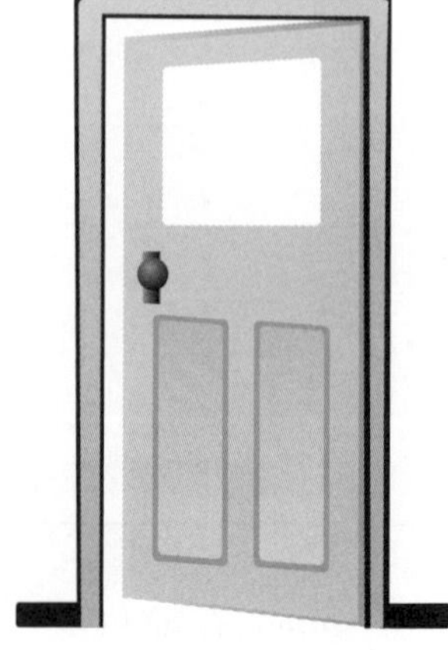

Number and Algebra

SET 3 Calculator division

Solve these divisions using a calculator.

1 $8\overline{)3608}$	**2** $9\overline{)4059}$	**3** $7\overline{)462}$
4 $8\overline{)364}$	**5** $8\overline{)4348}$	**6** $8\overline{)521}$
7 $4\overline{)3582}$	**8** $4\overline{)4813}$	**9** $6\overline{)4665}$
10 $3\overline{)2473}$	**11** $9\overline{)3877}$	**12** $6\overline{)1480}$
13 $15\overline{)390}$	**14** $17\overline{)595}$	**15** $26\overline{)936}$

Change these fractions to decimals.

16 $\frac{2}{5} = 0.$	**17** $\frac{2}{8} = 0.$	**18** $\frac{1}{4} = 0.$
19 $\frac{7}{10} = 0.$	**20** $\frac{9}{12} = 0.$	**21** $\frac{48}{100} = 0.$

SET 4 GST/money

To add GST to an item, add 10% ($\frac{1}{10}$).

Find the total cost of each manufactured item when the 10% GST is added.

1 $40

GST	
Total cost	

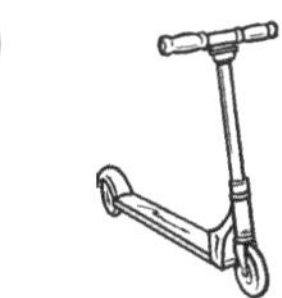

2 $90

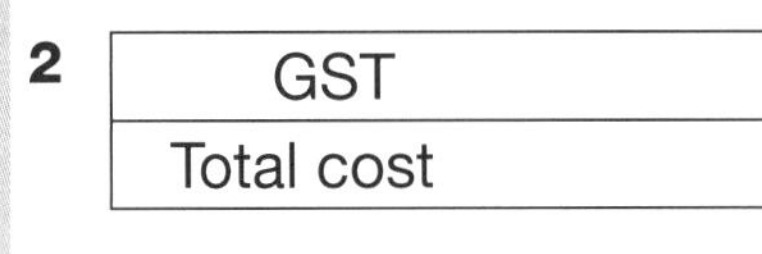

GST	
Total cost	

3 $60

GST	
Total cost	

4 $240

GST	
Total cost	

Measurement Timelines

Draw a line to match each invention to a point on the timeline.

1865 — 1880 — 1895 — 1910 — 1925 — 1940

Aeroplane	Automobile	Computer	Radio	Telephone	Television
1903	1885	1939	1895	1876	1926

Maths helpers

Length
10 millimetres (mm) = 1 centimetre (cm)
100 centimetres (cm) = 1 metre (m)
1000 metres (m) = 1 kilometre (km)

Mass
1000 grams (g) = 1 kilogram (kg)
1000 kilograms (kg) = 1 tonne (t)

Capacity
1000 millilitres (mL) = 1 litre (L)

Time
60 seconds = 1 minute
60 minutes = 1 hour
24 hours = 1 day
7 days = 1 week
14 days = 1 fortnight
12 months = 1 year
52 weeks = 1 year
365 days = 1 year
366 days = 1 leap year
10 years = 1 decade
100 years = 1 century

Months of the year
Thirty days has September, April, June and November. All the rest have thirty-one, except February alone, which has twenty-eight days clear and twenty-nine days each leap year.

Seasons
Summer: December, January, February
Autumn: March, April, May
Winter: June, July, August
Spring: September, October, November

Roman numerals
1 = I
2 = II
3 = III
4 = IV
5 = V
6 = VI
7 = VII
8 = VIII
9 = IX
10 = X
20 = XX
30 = XXX
40 = XL
50 = L
60 = LX
70 = LXX
80 = LXXX
90 = XC
100 = C
500 = D
1000 = M

Multiplication facts

×	0	1	2	3	4	5	6	7	8	9	10
0	0	0	0	0	0	0	0	0	0	0	0
1	0	1	2	3	4	5	6	7	8	9	10
2	0	2	4	6	8	10	12	14	16	18	20
3	0	3	6	9	12	15	18	21	24	27	30
4	0	4	8	12	16	20	24	28	32	36	40
5	0	5	10	15	20	25	30	35	40	45	50
6	0	6	12	18	24	30	36	42	48	54	60
7	0	7	14	21	28	35	42	49	56	63	70
8	0	8	16	24	32	40	48	56	64	72	80
9	0	9	18	27	36	45	54	63	72	81	90
10	0	10	20	30	40	50	60	70	80	90	100

Addition facts

+	2	3	4	5	6	7	8	9	10	11	12
2	4	5	6	7	8	9	10	11	12	13	14
3	5	6	7	8	9	10	11	12	13	14	15
4	6	7	8	9	10	11	12	13	14	15	16
5	7	8	9	10	11	12	13	14	15	16	17
6	8	9	10	11	12	13	14	15	16	17	18
7	9	10	11	12	13	14	15	16	17	18	19
8	10	11	12	13	14	15	16	17	18	19	20
9	11	12	13	14	15	16	17	18	19	20	21
10	12	13	14	15	16	17	18	19	20	21	22
11	13	14	15	16	17	18	19	20	21	22	23
12	14	15	16	17	18	19	20	21	22	23	24

Answers

UNIT 1 Number and Algebra

SET 1

1 12
2 6
3 15
4 6
5 16
6 25
7 36
8 $13.68
9 3
10 4
11 23
12 40
13 6
14 7000
15 $51

SET 2

1 6 × 4 = 24
4 × 6 = 24
24 ÷ 6 = 4
24 ÷ 4 = 6
2 8 × 3 = 24
3 × 8 = 24
24 ÷ 8 = 3
24 ÷ 3 = 8
3 7 × 5 = 35
5 × 7 = 35
35 ÷ 7 = 5
35 ÷ 5 = 7

SET 3

1 83
2 22
3 95
4 62
5 145
6 499
7 677
8 215
9 948
10 472

	Estimate	Answer
11	Hands on	81
12	Hands on	301
13	Hands on	260
14	Hands on	806
15	Hands on	601

SET 4

1 300
2 37
3 two hundred and thirty-two
4 Yes
5 $9 each
6 86 421
7 5
8 $14
9 97
10 16
11 Half
12 1, 2, 4, 7, 14, 28
13 $4.50
14 $45
15 Twenty-six thousand

Space

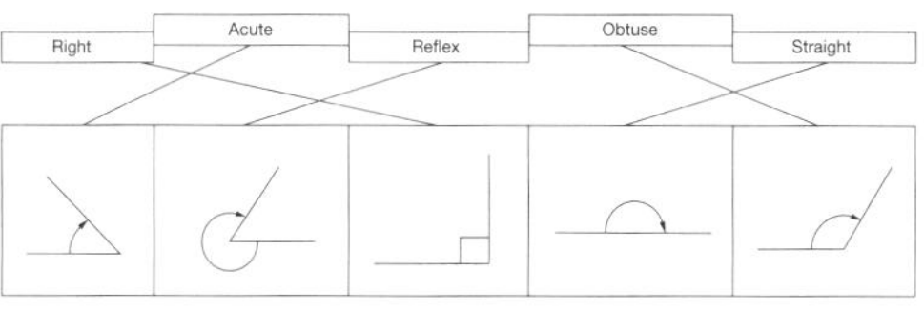

Measurement

1 mm/cm
2 mm
3 m/km
4 m
5 km
6 m

UNIT 2 Number and Algebra

SET 1

1 10
2 18
3 30
4 24
5 32
6 24
7 21
8 3215
9 2
10 3
11 55
12 40
13 10
14 6000
15 $35

SET 2

1 970
2 976
3 888
4 1034
5 995
6 1038
7 801
8 1412
9 $907

SET 3

1 $\frac{3}{6}$
2 $\frac{5}{6}$
3 $\frac{2}{3}$
4 $\frac{2}{6}$
5 $\frac{2}{3}$
6 $\frac{3}{6}$

Possible answers for 7–10:

7
8
9 
10
11 $\frac{1}{3}$
12 $\frac{3}{3}$
13 $\frac{5}{6}$

SET 4

1 18
2 90
3 $49
4 39 168
5 6
6 10
7 6
8 307, 332
9 $\frac{1}{2}$
10 14 679
11 $16 each
12 $21.60
13 1, 2, 4, 8, 16, 32
14 27
15 Thirty-nine thousand and six

Space

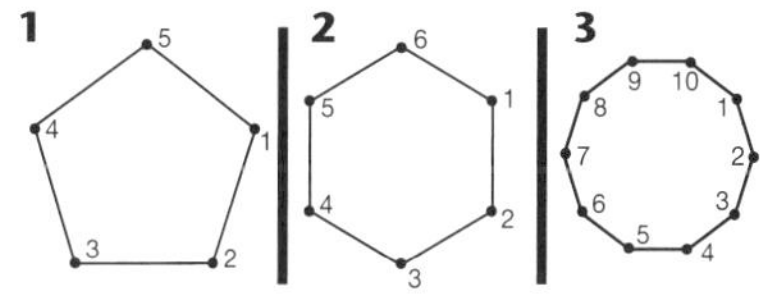

Measurement

	Length	Width	Area (L × W)
1	4 cm	2 cm	8 cm^2
2	3 cm	2 cm	6 cm^2
3	4 cm	3 cm	12 cm^2
4	7 cm	3 cm	21 cm^2

UNIT 3 Number and Algebra

SET 1

1 11
2 13
3 11
4 40
5 1
6 54
7 14
8 421
9 6
10 7
11 12
12 12
13 300
14 3
15 9

SET 2

1 201
2 $824
3 211
4 441
5 96
6 238
7 318
8 544
9 583
10 295
11 $262

SET 3

1 12 2 14
3 23 4 27
5 27 6 13
7 16 8 17
9 17 10 27
REALLY COOL
11 27 ÷ 9 = 3 and 27 ÷ 3 = 9
12 35 ÷ 5 = 7 and 35 ÷ 7 = 5
13 40 ÷ 8 = 5 and 40 ÷ 5 = 8
14 42 ÷ 7 = 6 and 42 ÷ 6 = 7
15 63 ÷ 9 = 7 and 63 ÷ 7 = 9
16 72 ÷ 9 = 8 and 72 ÷ 8 = 9

SET 4

1 280
2 $288
3 395 cm
4 25
5 145
6 474
7 189
8 $36.48
9 $1.55
10 80
11 104
12 $12.65
13 $27
14 5599
15 370
16

18	9	36	27	54	72	45	63
2	1	4	3	6	8	5	7

Statistics and Probability

1 Green $\frac{5}{20}$
2 Red $\frac{10}{20}$
3 Yellow $\frac{2}{20}$
4 Blue $\frac{3}{20}$

Measurement

1 24 cm^3
2 24 cm^3
3 21 cm^3
4 24 cm^3

Colour number 3

Answers

UNIT 4 Number and Algebra

SET 1

1 11
2 17
3 14
4 3
5 14
6 24
7 24
8 468
9 6
10 10
11 28
12 14
13 400
14 7
15 78

SET 2

1 15
2 150
3 1500
4 150
5 200
6 120
7 800
8 1600
9 50
10 600
11 86
12 24
13 87
14 38
15 87
16 83
17 15
18 76

SET 3

1 millions
2 thousands
3 tens
4 hundred thousands
5 hundreds
6 ten thousands
7 millions
8 ones
9 hundred thousands
10 hundreds
11 659 803, 1 030 584, 1 484 003
12 6 798 612, 9 365 804, 12 694 805
13 Two million, six hundred and five thousand, four hundred and twelve

SET 4

1 72
2 $6
3 $525
4 0.7
5 2 700
6 240
7 115
8 39
9 $47.50
10 56
11 119
12 $12.25
13 30
14 53
15 65
16 21, 34, 55, 89

Space

1 (1,4)
2 (5,3)
3 (3,2)
4 (1,1)
5 (5,1)

Measurement

Hands on.

UNIT 5 Number and Algebra

SET 1

1 13
2 13
3 1
4 3
5 40
6 54
7 32
8 671
9 6
10 7
11 56
12 19
13 5
14 30
15 7

SET 2

1 16, 160
2 24, 240
3 32, 320
4 40, 400
5 36, 360
6 28, 280
7 24, 240
8 36, 360
9 48, 480
10 60, 600
11 54, 540
12 42, 420
13 $300
14 $280
15 $270
16 $220

SET 3

1 59 km
2 102 km
3 341 km
4 384 km
5 282 km
6 43 km
7 No. $36 is the right change.

SET 4

1 2654
2 1350
3 97 631
4 $95.00
5 92
6 16th
7 $2.15
8 18
9 2023
10 $1.80
11 No
12 24
13 $39
14 103
15 46 398
16 Acute

Statistics and Probability

1 Red
2 Pink
3 Yes
4 Yes
5 Red

Space

1 Triangular prism
2 Square pyramid
3 Cylinder
4 Hexagonal prism
5 Hexagonal pyramid

UNIT 6 Number and Algebra

SET 1

1 12
2 13
3 5
4 2
5 24
6 40
7 27
8 680
9 7
10 8
11 36
12 20
13 7
14 4
15 $2.50

SET 2

1 18 r 2
2 16 r 2
3 14 r 2
4 15 r 3
5 15 r 4
6 12 r 4
7 11 r 3
8 12 r 3
9 12 r 1
10 15 r 1
11 14 r 1
12 24 r 2
13 14 r 1
14 9 r 1
15 7 r 6
16 $19
17 11 r 1
18 $9

SET 3

1 6, 12, 18, 24, 30, 36, 42
2 36
3 23, 27, 31, 35, 39, 43, 47
4 2, 7, 12, 17, 22, 27, 32
5 59, 56, 53, 50, 47, 44, 41
6 1, 2, 4, 8, 16, 32, 64

SET 4

1 About 1250
2 470
3 8500
4 16
5 42 668
6 25th
7 $1.90
8 72
9 126
10 $21
11 7 tenths + 3 hundredths
12 $48
13 $19.35
14 9000
15 56 969
16 5 to 9 or 8:55

Statistics and Probability

1 $6
2 No, only 1
3 $2, $7 and $9
4 $87 ($20 + $25 + $42)

Measurement

1 10:20 am
2 11:15 am
3 3:25 pm
4 9:50 pm
5 10:45 pm

UNIT 7 Number and Algebra

SET 1

1 42
2 40
3 12
4 38
5 27
6 21
7 5
8 6
9 80
10 35
11 5
12 60
13 80c
14 90 000
15 10

SET 2

1 108
2 390
3 296
4 200
5 402
6 147
7 570
8 $891

SET 3

Possible solutions, shading may vary.

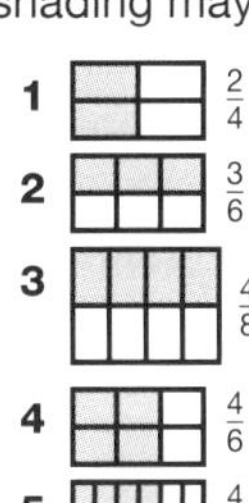

1 $\frac{2}{4}$
2 $\frac{3}{6}$
3 $\frac{4}{8}$
4 $\frac{4}{6}$
5 $\frac{4}{6}$
6 $\frac{1}{3}$, $\frac{2}{6}$
7 $\frac{2}{3}$, $\frac{4}{6}$
8 $\frac{1}{2}$, $\frac{4}{8}$
9 $\frac{2}{4}$, $\frac{1}{2}$
10 $\frac{4}{10}$, $\frac{2}{5}$
11 $\frac{3}{4}$, $\frac{6}{8}$

SET 4

1 0.81
2 19
3 19
4 80
5 20
6 2356, 4127, 26 007
7 40
8 Yes
9 $17.50
10 32
11 1425
12 1, 2, 4, 5, 8, 10, 20, 40
13 7
14 2.5
15 20 girls

Space

1 30°
2 90°
3 70°
4 120°
5 110°
6 40°
7 20°

Measurement

1 2000 m
2 km
2 4000 m
4 km

UNIT 8 Number and Algebra

SET 1

1 5
2 8
3 21
4 33
5 27
6 16
7 60
8 28
9 30
10 2607
11 35
12 63
13 4
14 15
15 12

SET 2

1 5839
2 9465
3 8633
4 9994
5 9063
6 8722
7 1285
8 3680
9 900
10 1800
11 5400
12 6700

SET 3

1 9
2 16
3 14
4 21
5 6
6 6
7 6
8 8
9 5
10 3
11 [10] (square)
12 (triangle 7) + (circle 12)
13 (circle 12) + [10] = 22
14 [10] + (circle 12) + (hexagon 3) = 25
15 (hexagon 3) × (circle 12) + [10] = 46
16 ([10] − (hexagon 3)) × (circle 12) = 84

SET 4

1 30
2 $45
3 True
4 1.0
5 4
6 22
7 165
8 23 467
9 Rectangles (includes squares)
10 154
11 No
12 4 am
13 $2
14 7:25
15 Thirty thousand and twenty-seven
16 114

Space

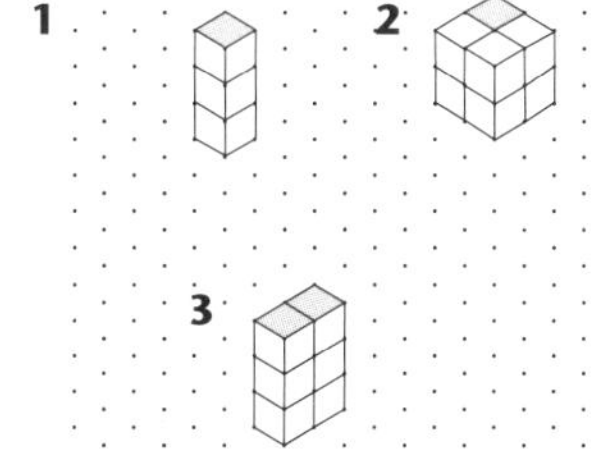

Statistics and Probability

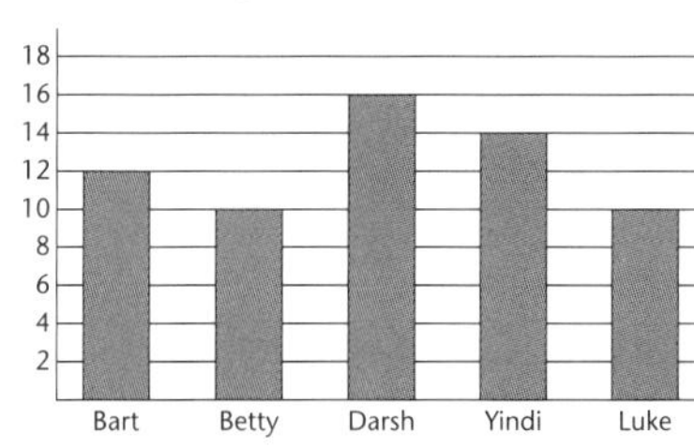

UNIT 9 Number and Algebra

SET 1

1 24
2 7
3 48
4 8
5 40
6 5
7 18
8 8
9 11
10 24
11 60
12 17
13 4
14 2
15 12

SET 2

1 60, 20, 80, 40, 100, 50, 90
2 20, 10
3 12, 6
4 32, 16
5 100, 50
6 44, 22
7 80, 40
8 28, 14, 7
9 40, 20, 10
10 100, 50, 25
11 80, 40, 20
12 48, 24, 12

SET 3

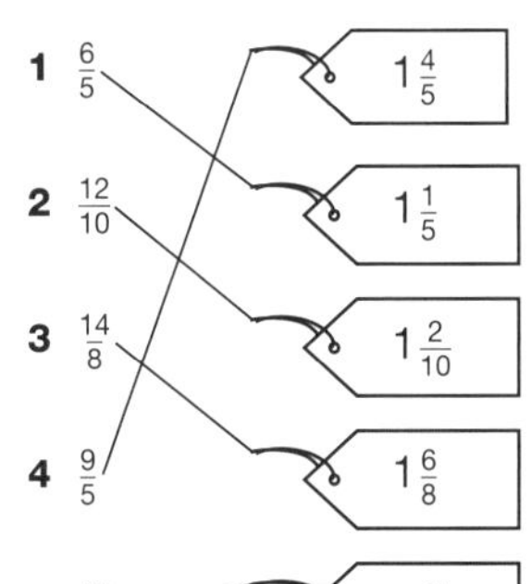
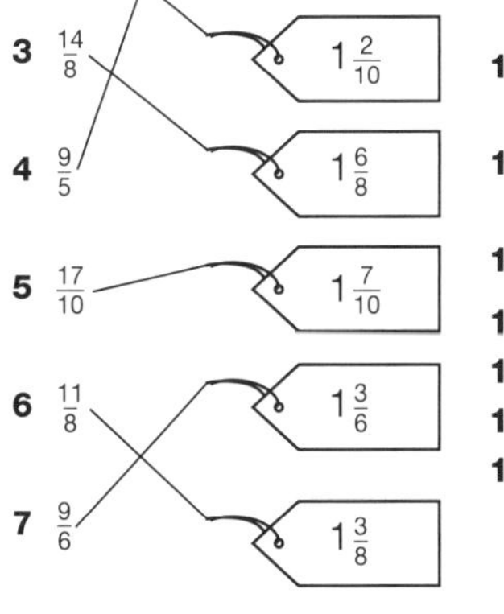

8 $\frac{11}{8} = 1\frac{3}{8}$
9 $\frac{7}{5} = 1\frac{2}{5}$
10 $\frac{16}{10} = 1\frac{6}{10}$
11 $\frac{8}{6} = 1\frac{2}{6}$
12 $\frac{10}{8} = 1\frac{2}{8}$
13 $\frac{18}{12} = 1\frac{6}{12}$
14 False
15 False
16 False
17 False

SET 4

1 About 700
2 13.8
3 750
4 $22.50
5 999 m
6 471
7 $360
8 14
9 1392, 1422
10 $18.50
11 $18.35
12 $377
13 7905
14 Anna: $30
Biyu: $15
Cath: $10
Deni: $5

Space

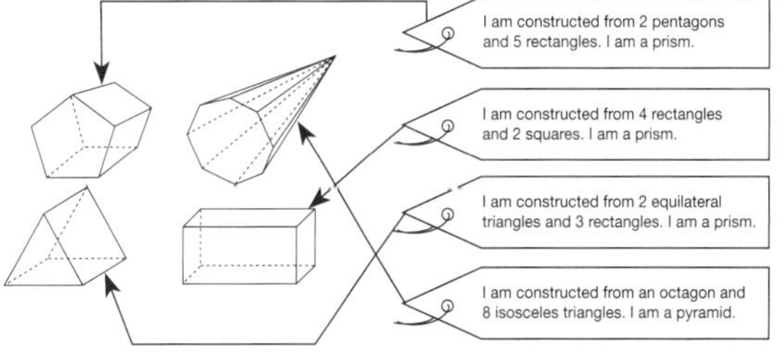

Statistics and Probability

1 6 2 5 3 3 4 5 5 41–50 kg

Answers

UNIT 10 Number and Algebra

SET 1

1 21
2 18
3 54
4 6
5 ÷
6 ÷
7 +
8 108
9 –
10 28
11 120 min
12 15c
13 50
14 1
15 96

SET 2

1 1561
2 3568
3 2080
4 5415
5 762
6 420
7 304
8 $6.45
9 308
10 $352

SET 3

1 Composite
2 Prime
3 Composite
4 Prime
5 Composite
6 Composite
7 Composite
8 Composite
9 Prime
10 Prime
11 Likely answer: 11 or 13
12 No
13 No
14 97
15 They are all multiples of 2.

SET 4

1 False
2 twenty-nine
3 9
4 1, 3, 5, 9, 15, 45
5 8
6 7
7 2700
8 No
9 $16
10 14
11 27
12 35 427, 28 963, 27 301
13 $\frac{3}{4}$
14 $2.25
15 Twenty thousand two hundred and six

Measurement

Hands on.

Statistics and Probability

1 True
2 False
3 True
4 False
5 True

UNIT 11 Number and Algebra

SET 1

1 22
2 +
3 15
4 –
5 35
6 ÷
7 8
8 ×
9 210
10 60
11 63
12 16
13 56c
14 100
15 210

SET 2

1 ✓$7 \times 3 = 21$
2 ✓$7 \times 7 = 49$
3 ✓$6 \times 9 = 54$
4 ✗$8 \times 7 = 56$
5 ✗$8 \times 4 = 32$
6 ✓$9 \times 9 = 81$
7 ✗$6 \times 8 = 48$
8 ✓$5 \times 9 = 45$
9 ✓$6 \times 7 = 42$
10 24
11 16
12 11 r 2
13 27
14 16 r 3
15 18
16 26
17 12 r 4
18 16 r 2

SET 3

1 0.55
2 0.69
3 0.33

Possible solutions:

4 0.25
5 0.79
6 0.10
7 0.80

SET 4

1 425
2 Yes
3 6
4 147
5 39
6 $11
7 296
8 $16
9 5
10 7496, 20 206, 53 224
11 42 cm
12 42
13 150 km

Number and Algebra

1 50
2 100
3 150
4 500
5 1000
6 1500
7 5000
8 10 000
9 15 000
10
5 | tens of thousands | 9 | thousands | 6 | hundreds | 3 | tens | 2 | ones
5 | 9 | thousands | 6 | hundreds | 3 | tens | 2 | ones
5 | 9 | 6 | hundreds | 3 | tens | 2 | ones
5 | 9 | 6 | 3 | tens | 2 | ones
5 | 9 | 6 | 3 | 2 | ones

Measurement

1 24 cm^2
2 12 cm^2
3 20 cm^2

UNIT 12 Number and Algebra

SET 1

1 44
2 $12
3 6 tens + 8 ones
4 $17
5 25
6 239
7 $7.70
8 49
9 $2.60
10 4000
11 9
12 30
13 $40
14 100
15 $7\frac{1}{2}$

SET 2

1

Taxi	$21 425
Sports car	$25 444
	$46 869

2

Truck	$42 600
2-door car	$19 999
	$62 599

3

Mini-bus	$35 500
Sports car	$25 444
	$60 944

4

2-door car	$19 999
Mini-van	$29 999
	$49 998

SET 3

1 $2.25
2 $3.60
3 $6.90
4 $9.60
5 $8.35
6 $11.10
7 $15
8 $21
9 $50
10 Yes

SET 4

1 56
2 $5.03
3 262
4 1431
5 68 cm
6 $15.22
7 7000
8 $3.60
9 $1.60
10 14, 21, 28, 35, 42, 49
11 $42.60
12 75
13 155
14 860
15 152 hundreds
16 20
17 About 43 km

Statistics and Probability

1 4 km
2 8 km
3 12 km
4 0
5 3 km per hour

Measurement

1 (30 cm^3)
2 (72 cm^3)

UNIT 13 Number and Algebra

SET 1

1 12
2 90
3 40
4 4
5 7
6 21
7 24
8 608
9 90
10 23
11 $7.70
12 10
13 621
14 $30
15 56

SET 2

1 1092
2 2480
3 2009
4 3045
5 2310
6 2760
7 1096
8 $1764

SET 3

1 24
2 23
3 10
4 11
5 5
6 6
7 8
8 10
9 10
10 5
11 18
12 7
13 9
14 15
15 11
16 3

SET 4

1 25 000
2 1829
3 1, 2, 3, 6, 7, 14, 21, 42
4 146
5 154
6 $39
7 0.7
8 38 000
9 10 × 9
10 4658
11 $2.50
12 7:10 pm
13 6 cm
14 $3.95 × 10 = $39.50

Space

1 **a** 110°
b 20°
c 90°
2

Measurement

1 0930
2 1200
3 1330
4 1600
5 1900
6 A Country Practice
7 Night Court

UNIT 14 Number and Algebra

SET 1

1 40
2 ÷
3 ×
4 445
5 $3.60
6 45
7 4
8 77
9 85
10 24
11 107
12 14
13 200
14 45
15 5

SET 2

1 2222
2 1132
3 2113
4 2112
5 2221
6 $3974
7 $2724
8 $1214

SET 3

1 7, 13, 19, 25, 31, 37, 43, 49
2 42, 47, 52, 57, 62, 67, 72, 77
3 6, 14, 22, 30, 38, 46, 54, 62
4 21, 20, 19, 18, 17, 16, 15, 14
5 9, 7, 5, 3

SET 4

1 19 000
2 99
3 $650
4 1, 2, 4, 8, 16, 32
5 $9\frac{1}{2}$
6 $2.55
7 40
8 80%
9 80c
10 6 cm
11 15
12 150
13 90c each
14 8
15 24
16 4 m × 4 m

Space

Hands on.

Measurement

1 18 cm
2 24 cm
3 26 cm

UNIT 15 Number and Algebra

SET 1

1 1305
2 21
3 +
4 27
5 −
6 54
7 6
8 ×
9 ÷
10 56
11 1000
12 25c
13 7000
14 10
15 161

SET 2

1 26 207
2 80 067
3 754 175
4 63 367
5 343 500
6 Twenty-six thousand and seven
7 Five hundred and ten thousand, one hundred and fifty
8 9 ten thousands
9 64 000
10 60 000
11 763 518
12 300 000 + 1000 + 500 + 7

SET 3

1 $\frac{1}{4}, \frac{1}{2}, \frac{3}{4}, 1, 1\frac{1}{4}, 1\frac{1}{2}, 1\frac{3}{4}, 2$
2 $\frac{1}{3}, \frac{2}{3}, 1, 1\frac{1}{3}, 1\frac{2}{3}, 2$
3 $\frac{1}{5}, \frac{2}{5}, \frac{3}{5}, \frac{4}{5}, 1, 1\frac{1}{5}, 1\frac{2}{5}$
4 $\frac{1}{8}, \frac{2}{8}, \frac{3}{8}, \frac{4}{8}, \frac{5}{8}, \frac{6}{8}, \frac{7}{8}, 1$
5 $\frac{1}{10}, \frac{2}{10}, \frac{3}{10}, \frac{4}{10}, \frac{5}{10}, \frac{6}{10}, \frac{7}{10}, \frac{8}{10}, \frac{9}{10}, 1$
6 $\frac{1}{2}, 1, 1\frac{1}{2}, 2, 2\frac{1}{2}$

SET 4

1 60
2 21 000
3 10 800
4 8500
5 27 226
6 6720
7 23, 29
8 2.5
9 215
10 $2\frac{1}{3}$
11 0.09
12 20 cm
13 20
14 40
15 $3.50
16 13 km

Statistics and Probability

1 3
2 4
3 2
4 3
5 Alphabetically according to surname

Measurement

1 8 m^2
2 16 m^2
3 28 m^2

Answers

UNIT 16 Number and Algebra

SET 1

1 12, 16
2 63
3 48
4 4
5 5
6 21
7 30
8 673
9 170
10 60
11 10
12 196
13 2000
14 \$3.60
15 \$6.60

SET 2

1 432
2 122
3 160 r 4
4 221 r 2
5 230
6 210
7 324
8 114
9 131
10 242
11 254
12 124
13 \$151.20
14 \$153

SET 3

(1-6 any four of the following answers)
1 1, 2, 4, 5, 10, 20
2 1, 2, 4, 8, 16
3 1, 2, 3, 4, 6, 12
4 1, 2, 4, 5, 8, 10, 20, 40
5 1, 2, 4, 8, 16, 32
6 1, 2, 3, 5, 6, 10, 15, 30
7 Yes
8 Yes
9 No
10 Yes
11 1, 2, 3, 4, 6, 8, 12, 24

SET 4

1 6 r 4
2 \$51 800
3 \$26.98
4 25 cm^2
5 246
6 \$70.24
7 425
8 40
9 Same
10 0.07
11 23.3
12 \$1.05
13 9 cm
14 37 942
15 23 632 or 32 623

Space

1 B and F
2 C
3 D and E
4 A
5 True
6 True

Space

1 A, B, C
2 C
3 D

UNIT 17 Number and Algebra

SET 1

1 5
2 4
3 +
4 12
5 38
6 14
7 ×
8 16
9 4
10 785
11 127
12 31
13 1
14 \$4.20
15 Hands on.

SET 2

1 \$80.49
2 \$46.88
3 Socks and shorts
4 \$26.75
5 Two caps and a pair of shorts

SET 3

1 $\frac{2}{4}$
2 $\frac{3}{10}$
3 $\frac{2}{3}$
4 $\frac{7}{12}$
5 $\frac{4}{5}$
6 $\frac{6}{10}$
7 $\frac{9}{10}$
8 $\frac{14}{100}$
9 $1\frac{2}{5}$
10 $2\frac{1}{5}$
11 $3\frac{7}{10}$
12 $4\frac{4}{10}$
13 $5\frac{5}{8}$

SET 4

1 516
2 33
3 230
4 2900
5 48 602, 37 100, 29 501
6 28 cm^2
7 52
8 150
9 \$9
10 632
11 20 000
12 16 r 1
13 3.30, 3.36
14 45
15 250 000
16 28 cm

Statistics and Probability

1 A 2 A 3 B 4 C
5 B and C

Space

T
F
S
BLUE
GREEN
RED

T
F
S
GREEN
BLUE
RED

UNIT 18 Number and Algebra

SET 1

1 54
2 6
3 17
4 5
5 ÷
6 +
7 –
8 ×
9 231
10 3
11 30 000
12 75c
13 8
14 54
15 3

SET 2

1 2472
2 2500
3 3033
4 1409
5 4745
6 1187
7 2500
8 6545
9 4835
10 4745

1 S	2 O	3 L	4 A	5 R

6 P	7 O	8 W	9 E	10 R

SET 3

1 15
2 18
3 19
4 24
5 30
6 31
7 20
8 15
9 9
10 26
11 T
12 F (48)
13 T
14 T
15 F (53)
16 T
17 F (9)
18 F (90)
19 F (300)
20 F (323)
21 Any pair that have a sum of 12 e.g., 12 + 0, 11 + 1, 10 + 2, 8 + 4

SET 4

1 29
2 79 000
3 7, 11, 13, 17, 19
4 4.17
5 800 g
6 134
7 37 200
8 162 cm
9 15
10 one hundred and forty-nine
11 $\frac{5}{10}$, $\frac{50}{100}$, $\frac{1}{2}$
12 \$1.35
13 135
14 35 256, 1479, 256
15 2000 g or 2 kg

Statistics and Probability

1 20°C
2 17°C
3 10°C
4 12°C
5 About 21°C

Measurement

1 4
2 2
3 8
4 1
5 5

UNIT 19 Number and Algebra

SET 1

1 25
2 14
3 48
4 5
5 18
6 19
7 415
8 195
9 24
10 22
11 17
12 63
13 21
14 28
15 $6

SET 2

[1] 1	2		[2] 2	[3] 4		[4] 1	[5] 9
7		[6] 1		[7] 6	5		4
	[8] 7	1				[9] 2	
[10] 1		[11] 1	[12] 2		[13] 1	8	4
[14] 2	9		0			7	
7		[15] 9	8				[16] 7
		3					4

SET 3

1 0.83 m
2 0.11 m
3 3.1 m
4 8.07 m
5 7.89 m
6 3.1, 3.2, 3.3, 3.4
7 8.12, 8.14, 8.16, 8.18
8 0.7, 0.75, 0.8, 0.85
9 3.25, 3.50, 3.75, 4
10 7.02, 7.09, 7.16, 7.23
11 0.78, 0.9, 1.02, 1.14
12 1.17, 1.37, 1.57, 1.77

SET 4

1 $37.50
2 64 962
3 42
4 243
5 10 × 10 × 10
6 11 000
7 About 700
8 9601
9 15
10 28.9
11 750
12 seven hundred and forty-five
13 $27
14 3.5
15 46 398
16 $67

Statistics and Probability

Aqua Swim School groups

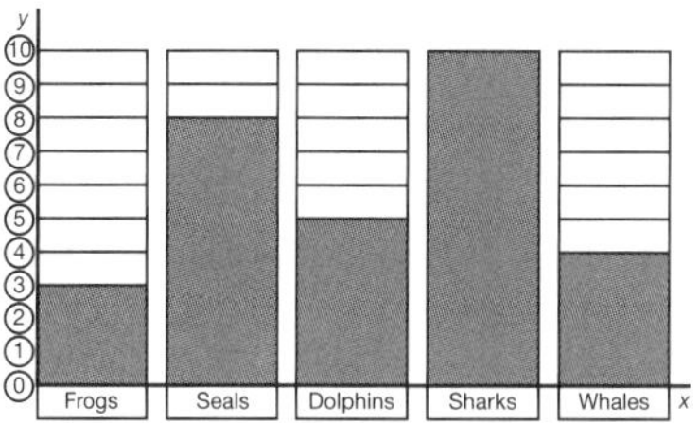

Measurement

	Less than 1 t	About 1 t	More than 1 t
elephant			✓
25 children		✓	
television	✓		
car		✓	
bulldozer			✓
refrigerator	✓		

UNIT 20 Number and Algebra

SET 1

1 20
2 12
3 35
4 5
5 14
6 371
7 144
8 24
9 28
10 33
11 72
12 81
13 7
14 48
15 144

SET 2

1 Reasonable
2 Reasonable
3 Reasonable
4 Reasonable
5 Reasonable
6 Reasonable
7 Reasonable
8 Unreasonable
9 Reasonable
10 Reasonable
11 29 753
12 27 431
13 49 574
14 27 844

SET 3

1 Students colour every shape except for the pentagon, hexagon and triangle.
2 Students colour the rhombus, second from the left.

SET 4

1 5.9
2 34 985
3 40
4 $3.60
5 4000
6 10 000
7 $1.54
8 3856
9 10:45
10 10 006
11 80
12 7.23
13 6200
14 $133
15 57 409
16 1925
17 Hanna is closest to the average of 150 cm.

Space

	Length	Width	Area
Carpet	8 m	6 m	48 m^2
Lino	4 m	3 m	12 m^2

Number and Algebra

1 $0, \frac{3}{10}, \frac{6}{10}, \frac{9}{10}, 1\frac{2}{10}, 1\frac{5}{10}, 1\frac{8}{10}, 2\frac{1}{10}$

2 $2, 1\frac{8}{10}, 1\frac{6}{10}, 1\frac{4}{10}, 1\frac{2}{10}, 1, \frac{8}{10}, \frac{6}{10}$

UNIT 21 Number and Algebra

SET 1

1 25
2 9
3 41
4 80
5 11
6 32
7 2 r 1
8 6000
9 +
10 –
11 38
12 48
13 27
14 $1 each
15 22

SET 2

1 $988
2 $1590
3 $2776
4 $2776
5 No
6 Brisbane
7 $5060

SET 3

1 $<$
2 $<$
3 $<$
4 $<$
5 $>$
6 $>$
7 $<$
8 =
9 $>$
10 $>$
11 9.514, 9.516
12 4.397, 4.399
13 6.798, 6.800
14 8.399, 8.401
15 3.600, 3.602

SET 4

1 950
2 4.07
3 4500
4 4.52
5 56 cm
6 $14
7 $16.50
8 8.32
9 9 m
10 No
11 Length: 48 m
Width: 48 m

Space

1 square
2 rectangle

Measurement

1 7:25 am
2 7:20 am
3 10:16 am
4 9:40 am
5 30 mins

Answers

UNIT 22 Number and Algebra

SET 1

1 ×
2 +
3 10 000
4 41
5 4
6 24
7 54
8 72
9 35
10 6
11 13
12 October
13 $9 each
14 1224
15 144

SET 2

1 $18 700
2 $26 850
3 $16 095
4 $10 360
5 $10 755
6 $26 455

SET 3

1 $\frac{9}{10}$
2 $\frac{3}{10}$
3 $\frac{5}{10}$
4 $\frac{4}{8}$
5 $\frac{5}{8}$
6 $\frac{3}{5}$
7 $\frac{4}{5}$
8 $\frac{2}{5}$
9 $\frac{5}{6}$
10 $\frac{2}{12}$
11 $\frac{5}{6}$
12 $\frac{11}{12}$
13 $\frac{4}{6}$
14 $\frac{7}{8}$
15 $\frac{6}{18}$
16 One person has $\frac{12}{18}$, the other person has $\frac{6}{18}$

SET 4

1 Yes
2 14.09
3 80
4 3.8
5 5 cm
6 $54
7 176
8 64 cm
9 0.34
10 860
11 $\frac{1}{4}$
12 47
13 $24
14 $25.90
15 60
16 $4.05

Statistics and Probability

1 Blue
2 15 had blue eyes
3 Yes
4 Brown

Space

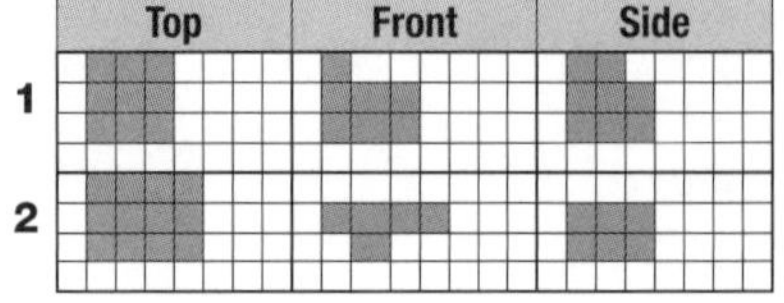

UNIT 23 Number and Algebra

SET 1

1 50
2 48
3 18
4 7
5 54
6 47
7 29
8 2
9 0
10 21
11 29
12 11
13 9
14 23
15 $70

SET 2

1 25 000
2 63 000
3 36 000
4 64 000
5 18 000
6 18 000
7 49 000
8 24 000
9 40 000
10 56 000
11 27 402
12 31 152
13 $38 270

SET 3

1 $\frac{1}{2}$
2 $\frac{2}{3}$
3 $\frac{3}{4}$
4 $\frac{4}{5}$
5 $\frac{5}{6}$
6 $\frac{7}{8}$
7 $\frac{11}{12}$
8 $1\frac{1}{2}$
9 $1\frac{2}{3}$
10 $1\frac{3}{4}$
11 $2\frac{1}{2}$
12 $2\frac{2}{3}$
13 $2\frac{3}{4}$
14 $3\frac{1}{2}$
15 $\frac{7}{8}$
16 $\frac{7}{8}$
17 $1\frac{6}{12}$
18 $\frac{4}{12}$
19 $\frac{2}{8}$
20 $\frac{1}{5}$
21 $\frac{2}{5}$

SET 4

1 $1\frac{2}{5}$
2 7000
3 2
4 5158
5 356
6 $22.40
7 45
8 $7.60
9 $18
10 11 pm
11 219
12 18 800
13 184
14 48 cm
15 12 629
16 64 kg

Space

8 7 6 5 4 3 2 1
A B C D E F G H

Statistics and Probability

Thick
Thin
Pepperoni
Chicken
Mushroom
Olives

UNIT 24 Number and Algebra

SET 1

1 5
2 36
3 46
4 42
5 37
6 32
7 32
8 +
9 ×
10 60
11 41
12 54
13 1, 2, 4, 5, 10, 20, 25, 50, 100
14 13 765
15 $12.50

SET 2

1 36
2 42
3 65
4 70
5 43 r 4
6 48
7 35 r 8
8 90
9 86 r 2
10 94
11 88
12 57 r 5
13 39
14 59 r 5
15 93 r 6
16 18
17 $86

SET 3

1 5
2 4
3 40
4 9
5 5
6 10
7 18
8 30
9 21
10 36
11 12
12 6
13 24
14 16
15 8
16 36
17 30
18 32

SET 4

1 925
2 48
3 $1.36
4 360°
5 False
6 48
7 119 cm
8 $29.50
9 90 cm
10 5852
11 $137.50
12 30 000
13 212
14 30
15 $72
16 2245

Space

Angle	Degrees
A	100°
B	30°
C	50°
D	130°

Measurement

Perth	Darwin	Brisbane	Sydney	Hobart	Melbourne	Adelaide
9 am	10:30 am	11 am	11 am	11 am	11 am	10:30 am
11:30 am	1 pm	1:30 pm	1:30 pm	1:30 pm	1:30 pm	1 pm
10:45 am	12:15 pm	12:45 pm	12:45 pm	12:45 pm	12:45 pm	12:15 pm

UNIT 25 Number and Algebra

SET 1

1 21
2 4
3 29
4 36
5 4
6 54
7 9
8 3000
9 ×
10 –
11 70
12 20
13 61
14 200
15 $210

SET 2

1 $9935
2 $196
3 $415
4 $2076
5 $208
6 35c

SET 3

	×	10	100	1000
1	2	20	200	2000
2	4	40	400	4000
3	8	80	800	8000
4	11	110	1100	11 000
5	12	120	1200	12 000

6 500
7 1020
8 1160
9 6150
10 8220
11 10 780
12 $9730

SET 4

1 30
2 27
3 True
4 20
5 180
6 570 km
7 5988
8 3.7
9 87
10 85 kg
11 Baseball: 20
Soccer: 30
Hockey: 10

Number and Algebra

1 $\frac{7}{5}$
2 $\frac{6}{5}$
3 $\frac{13}{10}$
4 $\frac{9}{8}$
5 $\frac{10}{8}$
6 $\frac{7}{5}$
7 $\frac{11}{10}$
8 $\frac{11}{10}$
9 $\frac{5}{4}$
10 $\frac{7}{6}$
11 $\frac{14}{10}$
12 $\frac{17}{10}$
13 $\frac{15}{10}$
14 $\frac{16}{10}$
15 $\frac{14}{10}$

Measurement

	Gross	Net	Difference
1	247 grams	200 grams	**47 grams**
2	**568 grams**	500 grams	68 grams
3	**172 grams**	100 grams	72 grams
4	325 grams	**250 grams**	125 grams
5	481 grams	**400 grams**	81 grams

UNIT 26 Number and Algebra

SET 1

1 24
2 31
3 9
4 56
5 30
6 14
7 19
8 5
9 5
10 18
11 $4.40
12 4
13 5
14 20
15 148

SET 2

1 528
2 1643
3 1350
4 891
5 1722
6 1890
7 324
8 682
9 6138
10 1008

SET 3

1 0.1
2 0.3
3 0.7
4 0.9
5 0.37
6 0.86
7 0.99
8 0.124
9 0.189
10 0.876
11 0.777
12 0.085
13 0.067
14 0.089
15 0.007
16 45.323, 45.324, 45.325
17 27.989, 27.990, 27.991
18 79.549, 79.550, 79.551
19 68.890, 68.891, 68.892
20 35.109, 35.110, 35.111

SET 4

1 12
2 28 cm
3 427
4 0.06
5 16 300
6 Tuesday 2 September
7 0
8 $225
9 3750
10 150
11 $20.10
12 $144
13 $15.00

Space

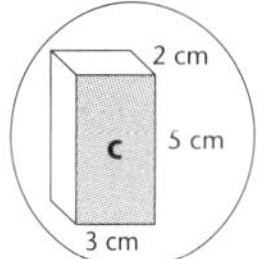

Statistics and Probability

1
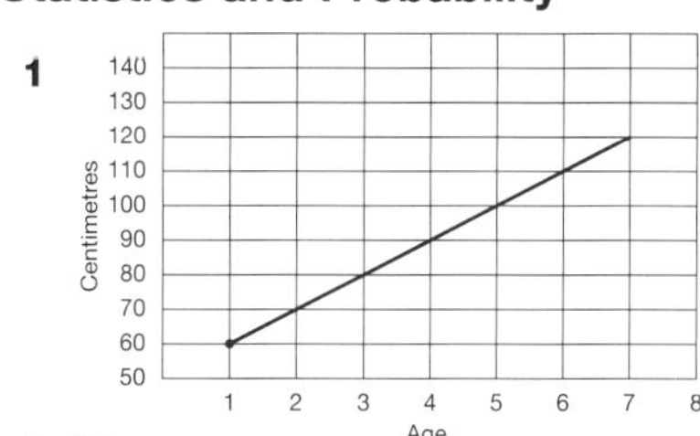

2 90 cm
3 115 cm
4 5 years

UNIT 27 Number and Algebra

SET 1

1 30
2 54
3 6
4 28
5 21
6 431
7 4000
8 $4 each
9 6
10 $2.40
11 180
12 13
13 36
14 3
15 $879

SET 2

1 9
2 7
3 4
4 6
5 7
6 9
7 7
8 50
9 5
10 7
11 12
12 8
13 6
14 4
15 3

SET 3

1 120
2 103
3 105
4 104
5 208 r 2
6 105
7 302 r 1
8 104
9 100
10 87 r 4
11 78
12 96
13 34 m^2
14 105
15 $101

SET 4

1 $24.50
2 9:20
3 17 700
4 60
5 5000
6 15.8
7 164
8 464
9 5
10 18
11 $14.40
12 45

13

Dice	1	2	3	4	5	6
1	2	3	4	5	6	7
2	3	4	5	6	7	8
3	4	5	6	7	8	9
4	5	6	7	8	9	10
5	6	7	8	9	10	11
6	7	8	9	10	11	12

7 is the most common score.

Measurement

Cubic centimetres	Millilitres
3 cm^3	3 mL
6 cm^3	6 mL
12 cm^3	12 mL
24 cm^3	24 mL
36 cm^3	36 mL
24 cm^3	24 mL

Cubic centimetres	Millilitres	Litres
200 cm^3	200 mL	
250 cm^3	250 mL	
400 cm^3	400 mL	
1000 cm^3	1000 mL	1 L
2000 cm^3	2000 mL	2 L
3000 cm^3	3000 mL	3 L

Measurement

Hands on.

Answers

UNIT 28 Number and Algebra

SET 1

1 28
2 $2.90
3 12 m
4 1, 2, 4, 8, 16
5 64th
6 94
7 $120
8 3
9 0
10 $6.50
11 548
12 0
13 34
14 7 tens, 2 ones
15 360

SET 2

1 28 000
2 81 000
3 48 925
4 36 608
5 23 442
6 28 137
7 62 006
8 18 612
9 14 356
10 55 992
11 60 464
12 56 592
13 $33 920

SET 3

1 742 064
2 962 613
3 959 312
4 888 147
5 993 235
6 979 918
7 757 779
8 869 194
9 $45 110

SET 4

1 0.6
2 14 cm
3 $6.25 each
4 7:09 pm
5 64 982
6 South-east
7 95c
8 11 October
9 5908
10 $9.90
11 1096
12 1680 km
13 $131
14 7

Space

1 **a** Palm Tree **b** Hilltop **c** Forestville **d** Dale
2 Possible answers:
Deep Lake = (12,6), (13,6), (14,6), (15,6), (12,5), (13,5), (14,5), (15,5), (14,4)
Dark Forest = (22,3), (22,4), (23,4), (21,5), (22,5), (23,5), (22,6), (23,6)

Statistics and Probability

1 St Joseph's
2 Mt Cook
3 Waverley
4 Bond St
5 English

UNIT 29 Number and Algebra

SET 1

1 5
2 50
3 8
4 72
5 27
6 74
7 32
8 5 r 1
9 600
10 ÷
11 +
12 $23
13 27
14 2
15 Thursday

SET 2

1 $11.72
2 $21.36
3 $30.21
4 $16.14
5 $79.57
6 $91.29

SET 3

1 $16.15
2 $14.30
3 $30.45
4 Glue sticks
5 **a** 33 g **b** $6

SET 4

1 27.43
2 $49
3 $57
4 39 cm
5 10
6 13.7
7 30
8 $\frac{1}{8}$
9 $42.50
10 20
11 4 metres

Statistics and Probability

1
2
3
4

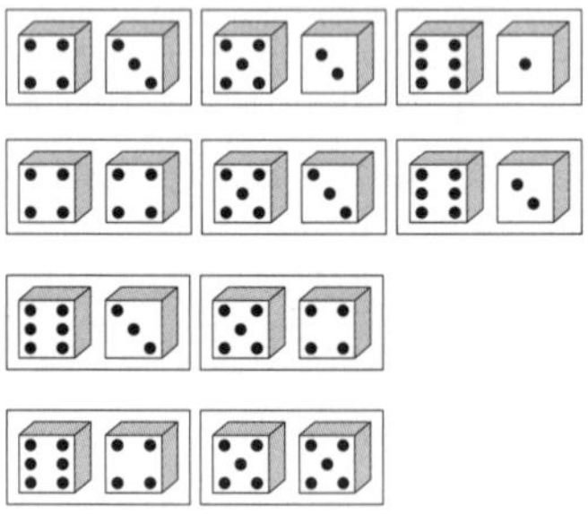

Space

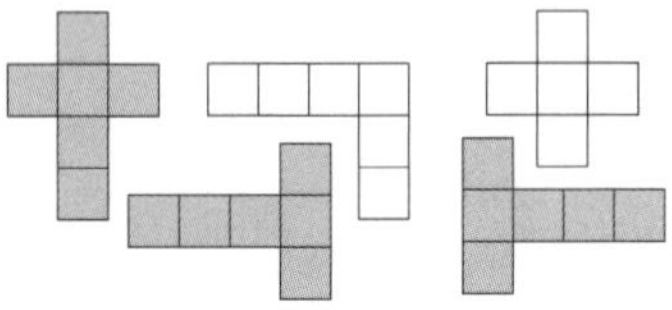

UNIT 30 Number and Algebra

SET 1

1 57
2 31
3 40
4 8
5 56
6 7
7 8
8 ÷
9 +
10 35
11 6
12 23
13 13 771
14 ninety
15 18

SET 2

1 $1015.38
2 $3205.95
3 617 517
4 808 747
5 628 199
6 501 894
7 Maitland and Port Macquarie

SET 3

1 1, 2, 5, 10
2 1, 2, 3, 4, 6, 12
3 1, 3, 5, 15
4 1, 2, 4, 8, 16
5 1, 2, 4, 5, 10, 20
6 1, 2, 3, 4, 6, 8, 12, 24
(Possible combinations)
7 23 x 12 = 23 x 4 x 3 = 276
8 $31 \times 15 = 31 \times 5 \times 3 = 465$
9 $41 \times 12 = 41 \times 2 \times 6 = 492$
10 $52 \times 15 = 52 \times 5 \times 3 = 780$
11 $43 \times 14 = 43 \times 2 \times 7 = 602$
12 $39 \times 12 = 39 \times 2 \times 6 = 468$
13 $40 \times 20 = 40 \times 5 \times 4 = 800$
14 $32 \times 16 = 32 \times 4 \times 4 = 512$
15 $44 \times 16 = 32 \times 8 \times 2 = 704$
16 $33 \times 20 = 33 \times 5 \times 4 = 660$

SET 4

1 16 cm^2
2 5
3 500 m
4 1179
5 70
6 $\frac{1}{4}$
7 37 400
8 170 cm
9 77
10 $11.60
11 17.17
12 $21.60
13 81
14 54
15 Straight (180°)
16 $45
17 $4800
18 About $240

Measurement

	Item	mm	cm	m	km	g	kg	mL	L	ha
1	The length of a car			✓						
2	The mass of a pencil					✓				
3	The capacity of a drum								✓	
4	The mass of a dog						✓			
5	The capacity of a small cup							✓		
6	The length of an ant	✓								
7	The length of a ruler		✓							
8	The area of a small farm									✓

Space

1 A 2 B 3 C 4 D 5 E

UNIT 31 Number and Algebra

SET 1

1 5
2 18
3 6 tens + 3 ones
4 345
5 32c
6 40
7 6
8 195
9 $110
10 $2
11 466
12 4
13 5 km
14 6
15 16

SET 2

1 $7
2 $9
3 $8
4 $12
5 $7
6 $3.50
7 3 kg $1.30
4 kg $1.20
5 kg $1.10

SET 3

1 ✓ 8 × 3 = 24
2 ✗ 7 × 7 = 49
3 ✗ 8 × 7 = 56
4 ✓ 8 × 8 = 64
5 ✓ 7 × 9 = 63
6 ✓ 23 + 77 = 100
7 ✓ 63 + 137 = 200
8 ✓ 45 + 48 = 93
9 ✗ 53 + 37 = 90
10 ✓ 837
11 ✓ 4623
12 ✗ 2792

SET 4

1 $1.05
2 2220
3 6
4 14 820
5 $2.50
6 $4.24
7 $1.50
8 72
9 20
10 $1\frac{1}{10}$
11 18 cm
12 36
13 60
14 800
15 84
16 4 (23, 29, 31, 37)

Measurement

	Area of ▭	Area of ◺
1	20 cm²	10 cm²
2	30 cm²	15 cm²
3	24 cm²	12 cm²
4	16 cm²	8 cm²
5	30 cm²	15 cm²

Measurement

1 False (square metres)
2 True
3 False (square metres)
4 True
5 True
6 False (square kilometres)

UNIT 32 Number and Algebra

SET 1

1 –
2 32
3 13
4 8
5 35
6 ÷
7 45
8 39
9 6
10 36 319
11 21
12 73
13 23
14 18
15 $18

SET 2

1

×	10	100
3	30	300
4	40	400
7	70	700
8	80	800
5	50	500

2

×	7	70
2	14	140
4	28	280
5	35	350
3	21	210
6	42	420

3 1799
4 19 524
5 40 544
6 945
7 2775
8 2064
9 $3096

SET 3

	Number	Nearest 1000	Nearest 100	Nearest 10
1	1089	1000	1100	1090
2	2189	2000	2200	2190
3	3588	4000	3600	3590
4	4768	5000	4800	4770
5	5217	5000	5200	5220

6 5 000 000

	Number	Rounded to	Divided by	Estimate	Answer
7	397	400	4	100	$99\frac{1}{4}$
8	789	800	8	100	$98\frac{5}{8}$
9	791	800	4	200	$197\frac{3}{4}$
10	149	150	5	30	$29\frac{4}{5}$
11	254	250	5	50	$50\frac{4}{5}$

SET 4

1 9 years
2 $\frac{3}{4}$
3 $\frac{1}{5}$
4 23 r 5
5 118 cm
6 192
7 125 cm
8 $68
9 $36.45
10 0.09
11 8 900
12 360°
13 27.03
14 280 cm²
15 True
16 67 (7 + 11 + 13 + 17 + 19)

Space

1 Newton Street
2 Adelaide Street
3 Cotter Street
4 Bow Street
5 Wow Street
6 Patrick Street

Measurement

1 Chad
2 Christina
3 Sarah
4 Alicia
5 Alicia
6 42 sec

UNIT 33 Number and Algebra

SET 1

1 41
2 20
3 ×
4 4 r 2
5 72
6 +
7 9
8 63
9 73
10 28
11 2000
12 15
13 6
14 1, 2, 3, 4, 6, 12
15 48

SET 2

1 $4
2 $3
3 $4
4 $3
5 $3
6 Fit Bodz $3
Gymbo $4

SET 3

	Fraction	Decimal	%
1	$\frac{10}{100}$	0.1	10%
2	$\frac{20}{100}$	0.2	20%
3	$\frac{25}{100}$	0.25	25%
4	$\frac{8}{10}$	0.8	80%
5	$\frac{1}{4}$	0.25	25%
6	$\frac{5}{10}$	0.5	50%
7	$\frac{75}{100}$	0.75	75%
8	0.04, $\frac{25}{100}$, 40%		
9	45%, $\frac{50}{100}$, 0.54		
10	0.06, 60%, $\frac{66}{100}$		
11	$\frac{40}{100}$, 0.45, 0.5		
12	5%, 0.1, $\frac{5}{10}$		
13	$\frac{10}{100}$, 11%, 0.7		
14	0.01, 10%, $\frac{11}{100}$		

SET 4

1 2700
2 Yes
3 725
4 $1\frac{2}{4}$ or $1\frac{1}{2}$
5 2452
6 23.7
7 $31.50
8 486
9 $189
10 44
11 1, 2, 3, 4, 6, 8, 12, 16, 24, 48
12 70 cm²
13 57
14 5
15 250
16 45

Statistics and Probability

1 No
2 No
3 Hands on.

Space

1

2

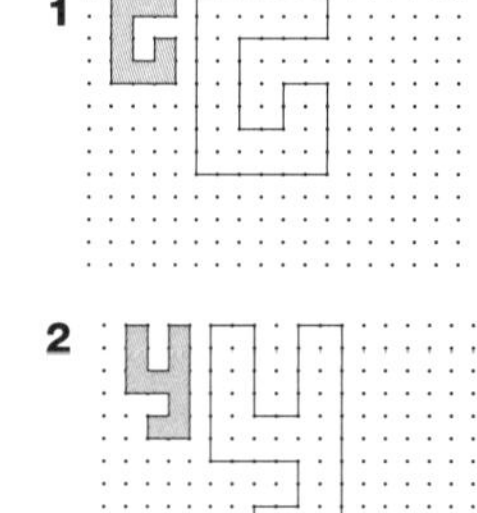

Answers

UNIT 34 Number and Algebra

SET 1

1 6
2 30
3 7 tens, 7 ones
4 465
5 54c
6 60
7 7
8 2272
9 $67.30
10 0
11 1648
12 7
13 9
14 $5
15 440

SET 2

1 $20
2 $40
3 $80
4 $100
5 $250
6 $30
7 $300
8 $400
9 $10
10 $5000
11 $50, $450
12 $70, $630
13 $200, $1800

SET 3

1 $9\frac{1}{2}$
2 $2\frac{3}{4}$
3 $5\frac{3}{10}$
4 $4\frac{2}{5}$
5 $9\frac{2}{4}$
6 $7\frac{2}{3}$
7 $6\frac{1}{3}$
8 $3\frac{5}{6}$

BRISBANE

9 $3\frac{1}{2}$ km

SET 4

1 70
2 24 cm
3 $240 each
4 14.8
5 117
6 104
7 $225
8 $16.15
9 $1.60
10 386
11 $68.25
12 15 000
13 588
14 127
15 0.6
16 2100

Space

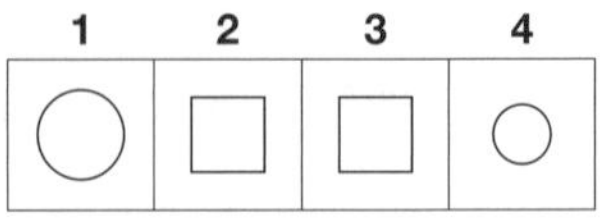

1	2	3	4
○	□	□	○

Statistics and Probability

1 Yes
2 No
3 12-year-olds
4 46

UNIT 35 Number and Algebra

SET 1

1 7
2 89
3 28
4 70
5 51
6 36
7 56
8 81
9 7000
10 ×
11 13
12 210
13 12 r 2
14 90 000 + 7000 + 6
15 20

SET 2

1 37
2 5
3 36
4 4
5 8
6 1
7 45
8 3
9 =
10 ≠
11 =
12 ≠
13 =
14 ≠
15 =

SET 3

1 451
2 451
3 66
4 45.5
5 543.5
6 65.125
7 895.5
8 1203.25
9 777.5
10 824.333
11 430.777
12 246.666
13 26
14 35
15 36
16 0.4
17 0.25
18 0.25
19 0.7
20 0.75
21 0.48

SET 4

1 $4 and $44
2 $9 and $99
3 $6 and $66
4 $24 and $264

Measurement

1 cm
2 mm
3 m
4 km
5 mL
6 cm^3
7 g
8 kg

Measurement

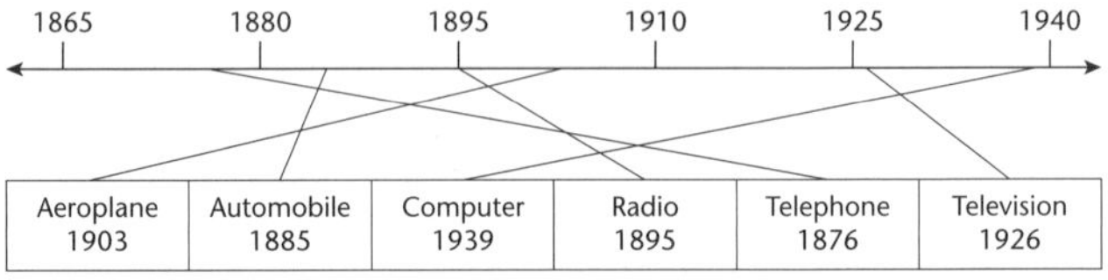

1865 1880 1895 1910 1925 1940

Aeroplane 1903	Automobile 1885	Computer 1939	Radio 1895	Telephone 1876	Television 1926